P9-APW-922

Microstructures of Surfaces

using Interferometry

Microstructures of Surfaces

using Interferometry

S. Tolansky, D.Sc., F.R.S.
Professor of Physics in the University of London,
Royal Holloway College

NEW YORK
American Elsevier Publishing Company, Inc.

First published 1968

American Elsevier Publishing Company, Inc.
52 Vanderbilt Avenue
New York, N.Y. 10017

Library of Congress Catalog Card Number: 68–55574

Printed in Great Britain by
William Clowes and Sons, Limited, London and Beccles

Preface

It is already twenty-three years since the multiple-beam interference method for revealing surface microtopography was first developed. Since then a large number of research reports devoted both to the exploitation of the technique and also to its theory and practice have appeared. Indeed, of such practical use is the method that two well-known instrument makers, the one in Sweden, the other in the U.S.A., are now marketing multiple-beam interference microscopes of especial use for metallurgical investigation and for the study of thin films.

However, to the writer, the special attraction of multiple-beam interferometry is its elegant simplicity of means, for indeed virtually nothing more is needed than a mercury lamp and a simple vacuum coating plant. Yet despite the simplicity of means it is possible to resolve in the up-down direction height changes of a mere five angstrom units, which is of molecular dimensions.

A realisation of the importance of a knowledge of surface microtopography is slowly beginning to be appreciated both by the pure scientist and the engineer. Today, so many engineering surfaces involve a smooth finish or are linked to problems of wear, that it is becoming more and more advantageous to know a good deal about the micro contours of surfaces. Without doubt interferometry gives the most satisfying picture of the shape contours of a surface. At best, expensive profilometers only show up line traces and even these are suspect if really soft surfaces are being traversed by a sharp diamond stylus which can well plough into the topography when making its record.

Here, in this atlas, a small selection has been chosen from the literally thousands of beautiful interferograms taken in the writer's laboratory over the years, including one taken when the technique was only just initiated. The selection is chosen merely to illustrate but a few of some of the regions studied. Although the technique sounds easy, in fact many subtle points are involved, discussion of which would be quite out of place in a text of this character. Thus the reader who might wish to its probe deeper is given suitable references on the last page of this monograph.

S. TOLANSKY

Acknowledgment

The following photographs contained in this atlas are reproduced by kind permission of Messrs. Longmans, Green & Co. Limited who first published them in *Surface Microtopography*, by S. Tolansky in 1960: Plates 10, 12, 13, 16, 17, 18, 19, 20, 22, 24, 25, 27, 28, 35, 36, 37, 38, 39, 40, 41, 42 and 49.

Contents

Chapter I

High Power Microscopy

Electron microscopy compared with interferometry

For studying the microstructures of surfaces the investigator is usually concerned with two, often related, but different aspects. These are high magnification and high resolution. The meaning of high magnification is self-evident, and it will be shown later that magnifications of even a million times are now obtainable by different devices. However a highly magnified image does not by any means necessarily mean that we have a highly resolved image. High resolution is just another way of describing the ability of a system to separate, clearly as individuals, details which are very close together. Unless details are separated, i.e. resolved, there is no point in boosting up the magnification further, for nothing is gained. Magnification increased beyond the point needed to see the limit resolved is called 'empty' magnification, and whilst it may sound imposing, scientifically it is of very little extra value to boost magnification to high empty figures.

If the optical system used has of itself a high resolving power, then a high magnification is needed to exploit this high resolution to full advantage. Indeed this is the only case in which high magnification is justified. In the ordinary optical microscope the limit of resolution is fixed by a number of factors, most important amongst which are the three now described. First, the resolution is proportional to the light-collecting *aperture* of the objective lens (the 'numerical aperture', as it is called). This is why, for high resolution, it is necessary to use a short focus lens with maximum possible ratio of width of lens to focal length. The resolution depends also on the *wavelength* of the light used, the shorter the wavelength the better. This is why it is possible to obtain a gain by a factor of about $\times 2$ in resolution by changing over from visible to ultra-violet light. Finally unless there is adequate *contrast* in the object, detail will not be visible, even if theoretically resolvable, and this is why the phase-contrast microscope has been invented, as a method of enhancing contrast to make theoretically optically resolvable detail visible to the eye.

The scientist has now available at least two distinctive microscopic techniques for producing both high magnification and high resolution. They are very different indeed and are in a sense complementary systems, giving quite different kinds of information. The one method is that of the *electron microscope*, the other that of *multiple-beam interferometry*, which is the subject of this monograph. The electron microscope is a refined, highly sophisticated piece of electronic engineering, full of clever complex gadgetry. A good instrument with attachments can easily cost £20,000, indeed the recent new million-volt electron microscope is likely to cost ten times this amount! The other system, multiple-beam interferometry, is a refined but quite inexpensive optical system which exploits the properties of light waves to advantage. It too requires skill and 'know-how', but uses little more than a conventional bench microscope.

The formidable electron microscope is capable of revealing the detailed structure of a very small region on an object, with magnification of maybe $\times 250{,}000$, and can resolve details as close together as 5 angstrom units (i.e. 1/50,000,000th inch) a *quantity smaller than many ordinary small molecules.*

Leaving out refinements, there are broadly speaking two ways of using an electron microscope. In the commoner method a very thin 'replica' is made of the surface to be studied by pouring on it, and allowing to set, a solution of a plastic such as Formvar. It has been established that when this dries into an exceedingly thin film, it moulds itself to the surface microstructure in a remarkable fashion, contouring in detail the surface microstructure, even almost down to molecular dimensions. According to where there are hills and hollows on the surface the replica is correspondingly thin and thick, locally. The replica is peeled off and placed within the evacuated chamber which is the main body of the electron microscope. On to the replica is directed an extremely narrow pencil beam of fast electrons. The replica is thin enough to pass the electrons, but the thicker parts, corresponding to the depressions on the original surface, scatter electrons away from the narrow beam. The electrons are focussed by magnetic lenses after passing the replica and these act as high magnifiers of the scatter pattern. The end result is that when the electron beam is finally focussed, after repeated magnification, on to a phosphorescent screen, a very highly magnified image of the structure on the replica is created.

The reason why the electron microscope is capable of high resolution is because fast moving electrons are optically the equivalent of extremely short wavelengths of light. Indeed the electrons used in a typical electron microscope carry with them waves which are perhaps 1/100,000 as small as visible light waves. It is because of this very short wavelength that the resolving power of the electron microscope is so high. It so happens that because of technical difficulties in magnetic lens design, the actual apertures of the lenses used in the electron microscope must be small. They need to be small also to sustain contrast, by avoiding collecting the scattered electrons coming from the thick regions of the replica. This considerably reduced lens aperture has two results. The one is that the resolution is by no means 100,000 times that of the optical microscope even though the wavelength is proportionately this amount shorter, for the optical microscope compensates a good deal with its bigger aperture. Still the electron microscope is certainly 500 times better, and in the best cases perhaps 1000 times better, in resolution than the best optical microscope. This is why high magnifications are justified.

The second result of the use of a small lens aperture is that the system has an unusually big 'depth of focus'. In any microscope, strictly speaking, there is sharp focus on a plane, but according to the lens aperture there is a tolerance above and below this plane in which the image is still in focus. This is called the depth of focus. For a high aperture lens this depth is very shallow and it is impossible to see clearly all parts of a thick object simultaneously. The smaller the lens aperture, the greater is the depth of focus and with a really small aperture both top and bottom of a relatively thick object appear at one and the same time in focus. Whilst this is at times a useful advantage, at other times it is a serious drawback. For suppose we seek information about the height or thickness of an object (up-down) as well as information about its size in extension (across) then we run into difficulty. We cannot assess height properly when all different heights are simultaneously in focus. One solution has been to use what is called 'shadow-casting' but this is only a partial remedy. The replica has heavy atoms beamed on it, directed at an inclined angle. The elevated regions cast shadows, just like the lengthened shadows cast by the sun when it is low on the horizon. These shadow lengths when measured give a figure for the height of the shadow-casting object. Clearly the method has limitations.

It is in connection with this measurement of up-down height that multiple-beam interferometry becomes appropriate as a complementary technique, as we shall show later.

SCANNING MICROSCOPE There does exist a modern modification of the electron microscope which avoids the use of a replica and which permits direct electron microscopic examination of solid bodies such as a piece of metal. This is called the scanning electron microscope. In this arrangement a tiny but intense micro-beam of electrons is imaged on the solid object. By the use of special coils this tiny spot is made to execute a 'raster', that is to say, just as on a television screen, the spot scans an area through a succession of close-packed linear tracks. Where the electrons strike the object, the object detail leads to emission of secondary electrons. These are fed on to a detector which then modulates a conventional television tube.

The system is capable of a very great range of magnification and although the resolving power is perhaps a hundred times less than that of the ordinary electron microscope, it exhibits such an astonishingly big depth of focus, that a three-dimensional image seems to appear. Again however, in spite of the brilliant photogenic character of the pictures obtained the question of the problem of measurement in depth still remains a difficulty.

Chapter II

Interference Fringes

Fringe Contours

When we drive through a hilly or undulating countryside we become conscious of the nature of the regional topography around us, the hills and dales, the valleys, the cols, perhaps the cliff ledges. Those of us who can read geographical contour maps, and who know that each contour line on the map is but the locus of all points which are at the same level above the sea, can with little difficulty recognise the character of a region from its map contour lines. A hill is shown by more or less circular rings of rising contours, and the more close-packed the contour lines the steeper is the hillside. It is typical practice on ordnance survey maps to use contour lines each representing height changes by 50 feet. Thus Plate 1 shows a contour map of part of the Prescelly hills in Cardiganshire. In the centre is a hill which rises up to 1750 feet, the north and south slopes being gentle, the eastern slope fairly steep. We shall find later that optical interference produces contour maps closely analogous to this pattern, but of course on a micro scale.

It happens that the majority of objects which surround us and which we imagine to be smooth and flat, such as for example a window pane, or a piece of highly polished silver plate, have in fact on a tiny scale a very complex microtopography. This is especially true of many objects of scientific and technological interest such as crystals, machined metals, plastics and so on. It is one purpose here to show that, on a micro scale, and perhaps encompassed within a height range too small to be detected even by a good quality high power optical microscope, there often exists on most objects a whole world of microtopographical detail, in many cases actually closely resembling a geographical contour map on a very reduced scale.

There can exist such a microstructure which is inherently natural to the object studied, but there can often also exist a major modification of this, introduced by some processing. With the aid of multiple-beam interferometry we shall create a truly three-dimensional contour map on selected objects. At the outset it is necessary to understand that this technique, which is comparable in power to the best electron microscope, is essentially powerful *only in the up-down direction*. This being so, the technique is exploited accordingly to give useful new information which can be profitable if topography in the up-down direction is needed. To take an example from geography, small changes in height may often be far more important than extensive changes in area. Thus some hundreds of square miles of Holland are a mere 10 feet below sea level, hence the need for the dykes. It is not so much the extensive area which interests the flood engineer, it is the mere 10 feet in depth that count so much. In like manner one can find on crystals areas which are relatively quite extensive, but which may be only a few molecules (indeed only one) high. It is the height of this molecular ledge which happens to be of interest, just as it was the 10 feet below sea level in Holland which mattered.

PLATE 1 Geographical contour map

Interferometry is the kind of approach which gives an answer to the height or depth of micro-ridges and ruts even when only of molecular dimensions, provided the area over which the object of study extends is amenable to be recognised and established by an ordinary microscope, working at any magnification less than $\times 1000$.

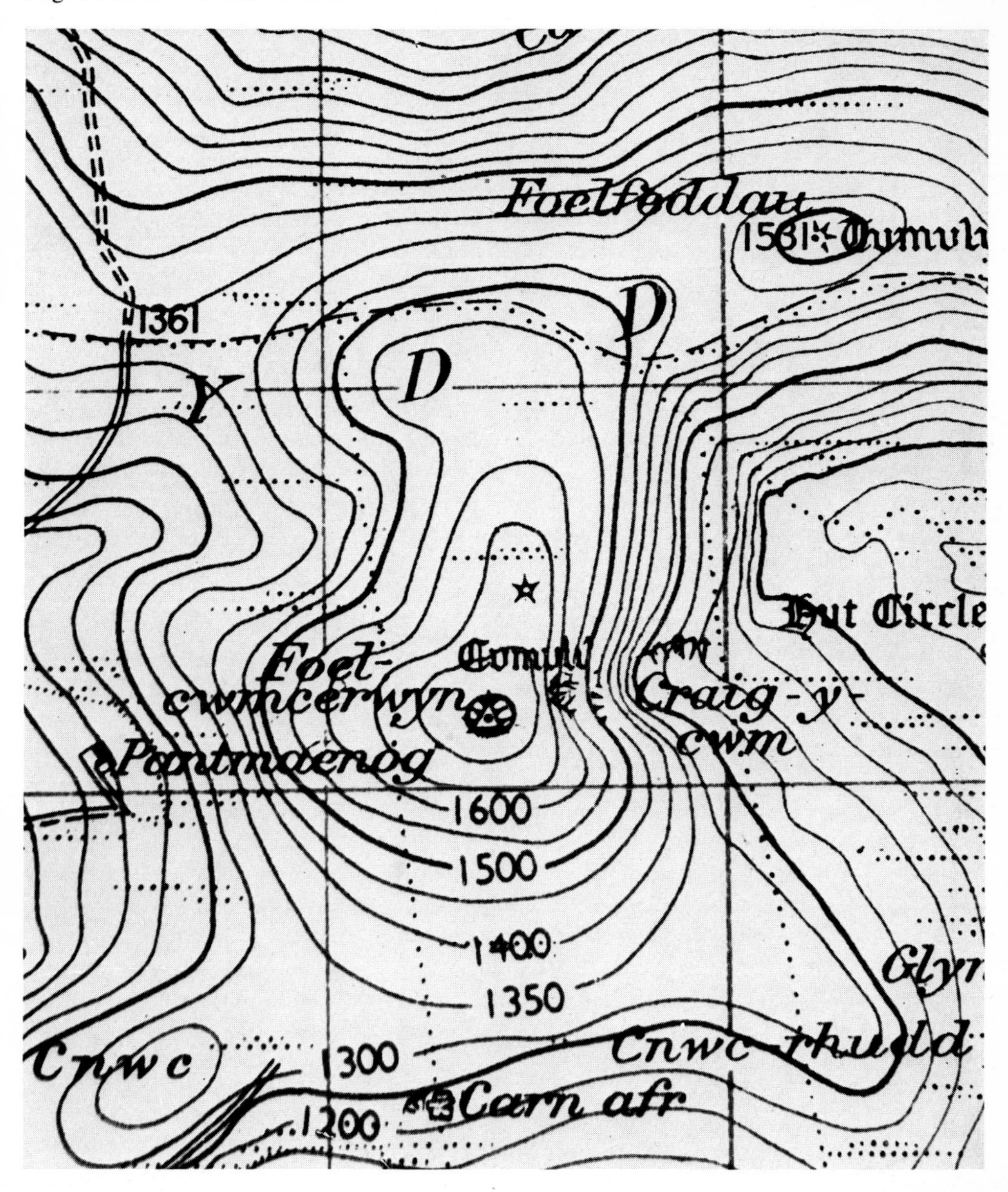

Multiple-beam interference

What happens in multiple-beam interferometry is that, by means of special optical devices, we see an object through the microscope and it *appears to be covered with optical contour lines*. These are similar in appearance and in significance to geographical contour lines, apart from the scale. For the optical contour lines are created through the interference of light waves reflected between the object under study and a flat plane surface placed close to it. On moving from one contour line to its nearest neighbour the surface height between the object and the reference plane has changed by half a light wave, which is about 1/100,000th inch. This optical scale is therefore 60,000,000 times as great as that on the 50 foot ordnance map contours.

When the optical interference technique is correctly used the optical contour lines (called *fringes*) are very sharp narrow lines which so closely follow micro features that it is possible to see and to measure small displacements which are no more than 1/500th part of the distance between adjacent fringes. This corresponds to 5 angstroms i.e. to 1/50,000,000th inch; indeed a quantity as small as the best resolution attainable by a high quality, costly electron microscope, but of course here the quantity revealed is in the up-down direction *only*. Across the object the magnification is merely that of the microscope used. It may be ×1000; then again it may be only ×10. It is easily possible to have fringes photographed on a print so that they are as much as 10 inches apart on this print. But in reality the true surface height change corresponding to this is only 1/100,000th inch. This means that the up-down magnification is actually ×1,000,000. If in this instance the magnification *across* is a mere ×10 then we have the curious situation in which the up-down scale is as much as 100,000 times that of the scale across the object. Admittedly this is exceptional, but in practice scale ratios of 1000:1 are not uncommon. This fact must constantly be remembered when interpreting *interferograms*, as they are called.

Because the fringes offer considerable resolution and magnification in the up-down direction, we find surfaces which appear smooth and glossy to a critical eye or to a good microscope, yet by interferometry can exhibit quite a fiercely complex microstructure of hills and dales and other features. For instance if a surface be lightly scratched or gently indented with a fine point we really plough up tiny ruts and micro-hillocks which can easily be studied by interferometry. If a surface be etched or corroded by chemical action, the local minute pits and changes in surface contour can be revealed. If 'technological' surfaces such as engine bearings or pistons are subject to wear, the microstructure interferometric changes can be revealed. Even the local thickness change produced on the edge of a razor blade by shaving can be detected. These and numerous other aspects all become fertile fields of study with this technique. Not only the engineer but the biologist too can exploit various forms of interferometry. The micro-shapes of some seeds, the topographical surface characters of certain fruits, the wear of the dentine on teeth, the structure of nail and horn, the character of certain sea shells; these are typical. The physicist, the chemist, the crystallographer, the metallurgist and the engineer can all variously find subjects which respond successfully to interferometric investigations and reveal unknown features. Some of these we will explore later.

FIGURE 1 Effect of reflectivity on fringe shape

Chapter III

Experimental Technique

Production of Fringes

To explain briefly how multiple beam fringes are produced the simplest possible case will be selected, the production of fringes between two nearly plane, rectangular pieces of glass, touching along an edge and slightly inclined to form a small-angled air wedge. Several separate optical conditions require to be satisfied before sharp narrow multiple-beam fringes can be produced.

To begin with, the surfaces must be coated with a layer which is (a) highly reflecting (b) slightly transmitting and (c) faithfully contours any microtopography. Experience has shown that a critically correct thickness of silver applied by deposition in a vacuum, through evaporation from a hot filament, meets all requirements. The thin film must be about 700 angstroms thick i.e. 1/350,000th inch and very little deviation from this is acceptable. Such a film, if well made under high purity and good vacuum conditions, will have a reflectivity of 95 per cent, absorbs about 4 per cent and transmits about 1 per cent. Furthermore, it has been established by exhaustive tests that such a silver film, when deposited correctly, has an astonishing fidelity in perfection of contour, moulding itself to microstructure detail down to molecular dimensions.

When two such silvered surfaces are placed close together, slightly inclined, and then illuminated at normal incidence with parallel monochromatic light, interference fringes form. Because of the high reflecting layers, when a beam enters the space between the two silvered surfaces, it bounces back and forth, a small fraction escaping each time. If (and this is a most important essential) the two surfaces are very close together indeed, *nowhere separated than a few light wavelengths*, then when the repeatedly reflected beams, i.e. the multiple beams, combine together, the optical result of summing up many beams is to sharpen up the fringes, making them appear like sharp narrow lines, instead of broad bands. The higher the reflectivity of the silver, the sharper and narrower become the fringes. Indeed a well developed silvering technique is an essential basic 'know-how' which must be learnt by experience in the laboratory.

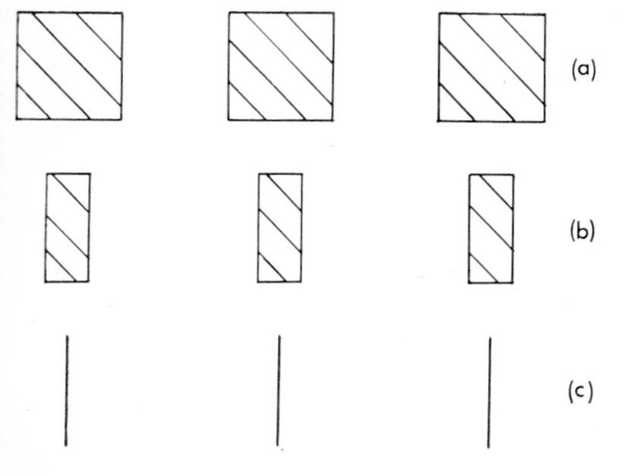

Fig. 1 is a schematic diagram of the end effect of increased reflectivity. In Fig 1(a), the appearance of fringes, when there is no silver on the plates, is shown. Light and dark regions are of equal width. Fig. 1(b) shows the kind of sharpening up taking place when a silver film of perhaps 70 per cent reflectivity is used. Fig. 1(c) shows the very sharp fringes produced when reflectivity is about 95 per cent. This is the desired situation.

Scale of fringes

Most of the fringe patterns shown in this monograph have been taken with the green line given out by a mercury arc spectrum. This is easily isolated by filters, has a wavelength of 5460 angstroms, i.e. 1/46,000th inch.

For our purpose it will be near enough, in round numbers to call this 1/50,000th inch.

It will be a convenience to summarise here the units we shall often use, which will be as follows.

For microscopically small quantities it is usual to use as a unit the micron (written μ) which is 1/1000 mm. The wavelength of green light can be taken to be more or less half a micron. It is usual to give wavelengths (written λ) in angstrom units, of which there are 10,000 in a micron. The wavelength of green light is 5000 angstrom units (written Å). Thus a height change corresponding to one fringe (half a wavelength) can be equated in approximate round numbers as:

$$1 \text{ fringe} = \lambda/2 = 2500 \text{ Å} = \frac{1}{40{,}000}\text{th cm} = \frac{1}{100{,}000}\text{th inch}$$

Since we will find that we can just measure displacements which are 1/500th of a fringe, the practical *limit* of measurement becomes

$$\frac{1}{500}\text{th fringe} = \frac{\lambda}{1000} = 5 \text{ Å} = \frac{1}{20{,}000{,}000}\text{th cm} = \frac{1}{50{,}000{,}000}\text{th inch}$$

To secure the optical contour pattern of a desired topography whether simple or complex, it is necessary to match that surface with some reference plane. We shall see that very fine scale structures indeed are revealed, so at once the question can be asked 'What is the reference plane and how do we know that it is free from any structure of its own which might falsify the result?' To this we have found a curious but adequate answer. It is more important in practice to have a 'smooth' reference surface, rather than a perfectly flat one. Some slight curvature can be tolerated provided the surface is so smooth as not to affect any microstructures under study. Indeed if we are looking only at a small area it is relatively easy to find an object reasonably flat over this small area. More important is the smoothness criterion. The solution to this problem was found over twenty years ago. It was discovered that cheap fire-polished glass, i.e. *flame-polished*, but not mechanically polished, can have a superlative degree of smoothness. It is as if the glass, a super-cooled liquid, takes on that natural smooth finish characteristic of a viscous liquid. For can anything be much smoother than a ripple-free surface of a highly viscous liquid? The proof of the perfection of this kind of reference surface is secured by the simple device of matching two such selected pieces of cheap fire-polished glass one against the other, after first covering correctly with silver, thus examining the fringes by multiple-beam interference methods. If two pieces of correctly silvered glass, say one inch squares, are matched to form a slight wedge, and if they are theoretically perfectly plane, they will give equidistant straight line fringes which will be quite free from structure of any kind. Suppose the two plates are set at a slight wedge angle, as in Fig. 2(a), then straight-line, parallel, equidistant fringes appear, parallel to the edge of contact; for at successive equal distances where space heights are $\lambda/2$, $2\lambda/2$, $3\lambda/2$ etc., fringes will form. The actual separation of the fringes, what is called the *dispersion*, is determined by the angle of the wedge. With a big angle they pack close together (Fig. 2(b)) and *vice versa* they open out as the angle is reduced.

(a)

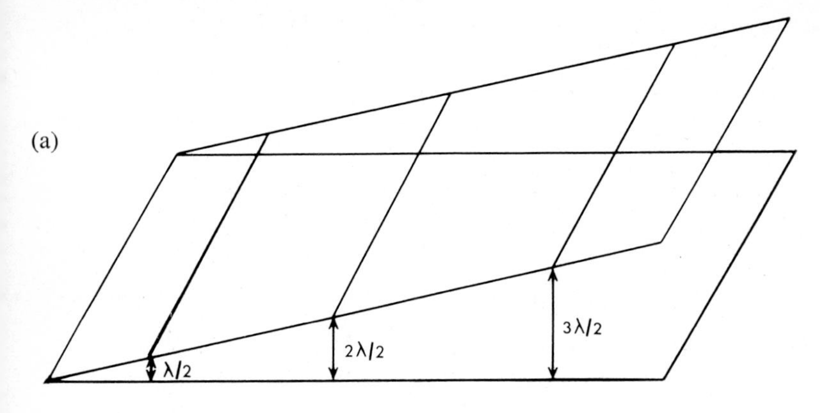

FIGURE 2 Straight line fringes formed by a wedge. (a) Wedge angle small (b) Wedge angle increased

(b)

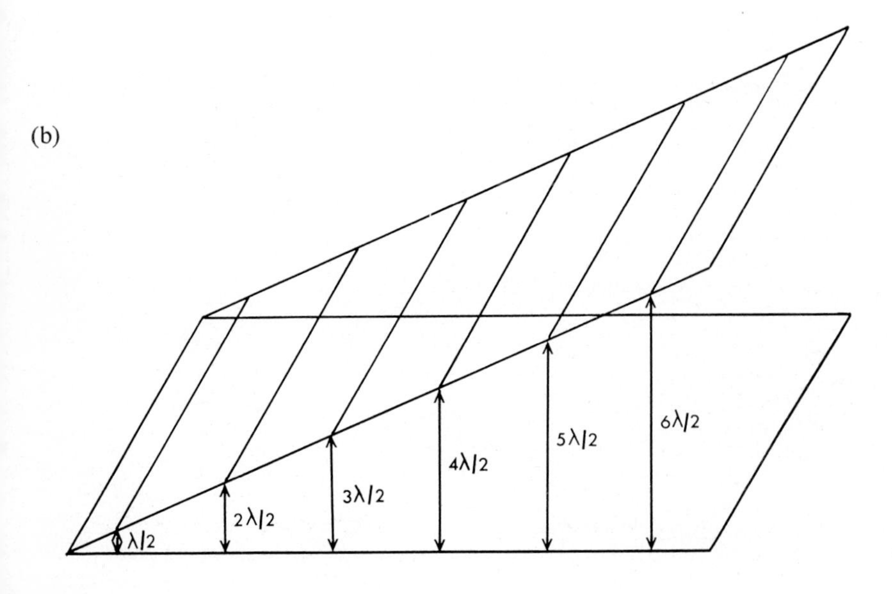

Glass surfaces

Let us now look at two typical kinds of different glass. Plate 2 shows the sharp multiple-beam interference fringe pattern given by two selected pieces of cheap fire-polished glass. There are several features of interest. To begin with the fringes are straight and smooth and show no structure. Secondly, they are almost perfectly parallel straight lines, so the surfaces over this region are adequately flat as well as smooth. Thirdly, the fringes are narrow bright lines on a broad dark background (we shall show later that it is possible to do better in this respect). Fourthly, the fringes on the print are $1\frac{1}{2}$ inches apart, but the real height change in moving from fringe to fringe is 1/100,000th inch; thus the effective up-down magnification is $\times$ 150,000 on this print.

This interferogram shows that we can with considerable confidence use such pieces of glass as reference 'flats' for studying other objects.

Now let us match one of these pieces of silvered fire-polished glass flats against a silvered rough irregular piece of glass. An interferogram ($\times$ 25) is shown in Plate 3. This reveals that the glass is not flat, but cylindrically curved. By auxiliary optical devices it is always possible to decide whether we are looking at a hill or a hollow. Here a ridge runs across almost two fringes in height (i.e. one light wave = 1/50,000th inch). Furthermore very small-scale minor irregularities are visible. In addition, since fringe separation in the ridge approximates to $2\frac{1}{2}$ inches, the true magnification in height is actually $\times$ 250,000. Here then is a case (a) in which magnification is high (b) resolution very good and (c) the ratio of the vertical to the horizontal scale is 10,000:1.

It will be noticed here that, while the fringes in Plate 2 are bright lines on a dark ground, those in Plate 3 are reversed, i.e. dark fringes on a bright ground. This is because, as will be explained shortly, Plate 2 was taken with transmitted light and Plate 3 with reflected light.

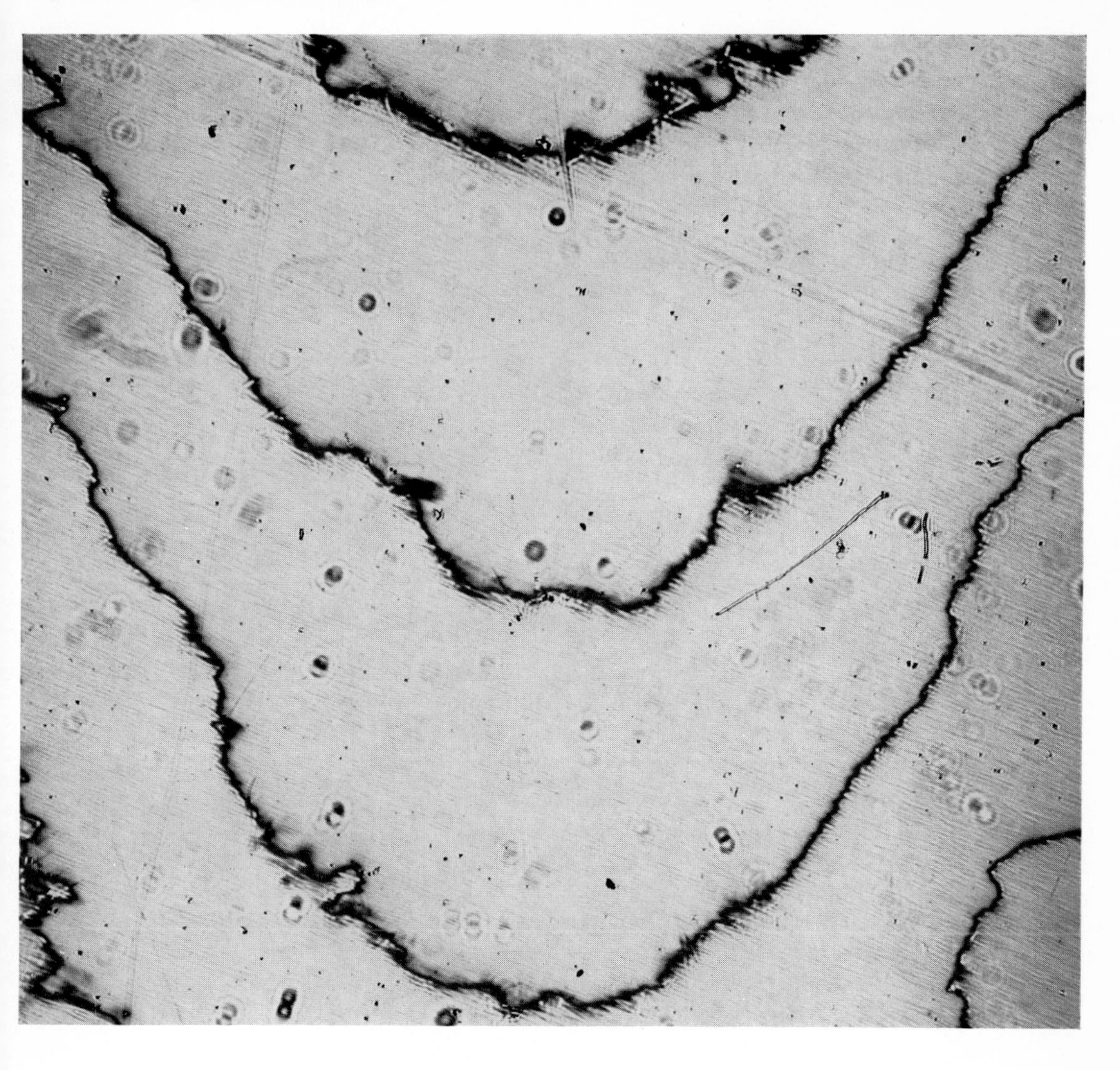

PLATE 3 Fringe pattern given by an irregular glass surface. Fringes with reflected light

PLATE 2 Smooth fringes given by two plates of fire-polished glass. Fringes with transmitted light

Experimental arrangements

Before going on to detailed studies we shall in a schematic way show in Fig. 3 how the microscope is used to view fringe patterns for a transparent object such as glass.

A is the mercury arc source, *B* the green filter. The small aperture *C* is brightly illuminated and is at the focus of the lens *D* which projects a parallel beam of light perpendicularly on the interference wedge *E*. The object is viewed with a microscope *F*, and under these conditions the fringes are seen In practice microscope powers may vary from say ×10 to say ×1000.

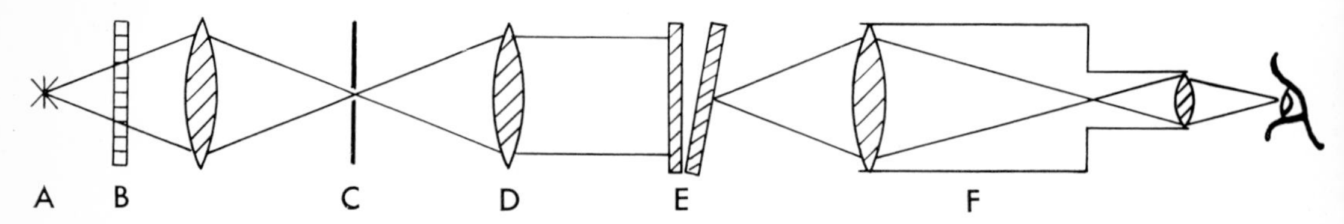

FIGURE 3 Arrangement for viewing fringes with a microscope

Now it so happens that very often it is desirable to study an opaque object such as a metal, in which case it is only possible to see fringes by reflected light, since clearly no light can go through the metal object! In fact it is often useful to use reflected fringes even when examining transparent objects; already in Plate 3 we have met such a case.

The use of reflection alters the character of the fringes and of course necessitates somewhat modified experimental arrangements, compared with those used for transmitted light. To illustrate in a simple fashion the difference reflected systems introduce, consider Fig. 4 in which we have in Fig. (a) two silvered partially transparent glass surfaces and in Fig. (b) one partially transparent silvered glass surface resting on a high reflecting opaque metal. In order to simplify the diagram, for illustration, we show the case where the surfaces are parallel and the light incidence is not quite normal. Without going into detail, it can be shown that the optical situation is such that there are two complementary systems. In (a) the fringes can be like those in Fig. 5(a) i.e. narrow bright lines (broadened here for clarity) on a dark background, such as the fringes in Plate 2. But in reflection the arrangement appears as in 5(b), narrow dark fringes on a bright background, like the fringes of Plate 3. To see reflected fringes with a microscope the optical arrangement needed is that in Fig. 6.

A source of light *A* is imaged by lens *B* on an aperture *D*, after passing a monochromatising filter *C*. Lens *E*, through the partial reflector *F*, forms a point image *G*, which, being at the focus of the microscope lens *H*, results in illumination of the interference object *O*, with parallel light at normal incidence. At the same time *O* is also at the ordinary position for microscopic image formation by lens *H*, which thus has the dual function of illuminating *O* with parallel light and at the same time giving a microscope image of the surface contoured with interference fringes. Thus it is that opaque objects can be studied by placing on them partially transparent high reflecting flats. If high microscope power is needed the flat must be very thin. In practice selected fire-polished glass cover slips are used, suitably coated with silver or with some other reflector.

When examining a surface, either by combinations of several wavelengths, or by using convergent light, or by applying slight pressure, it is always possible to make sure that a given group of fringes refers to an *up* or a *down* feature. If there is any doubt it can be settled by the use of an auxiliary type of fringe using white light.

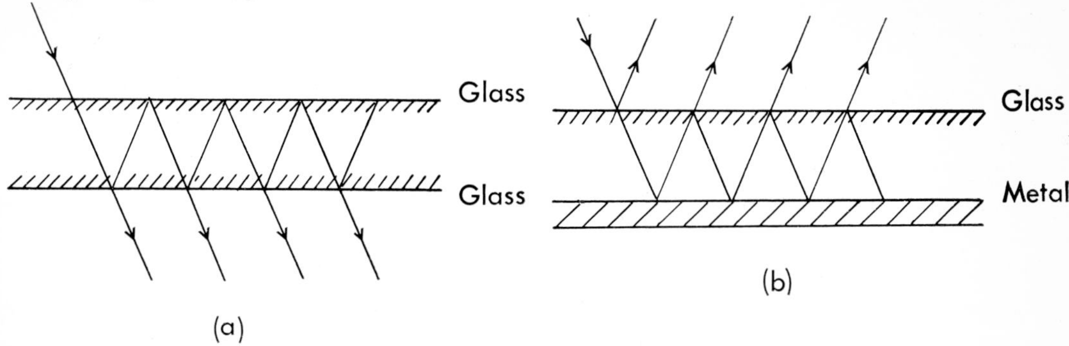

FIGURE 4 Light beam paths. (a) Transmission (b) Reflection

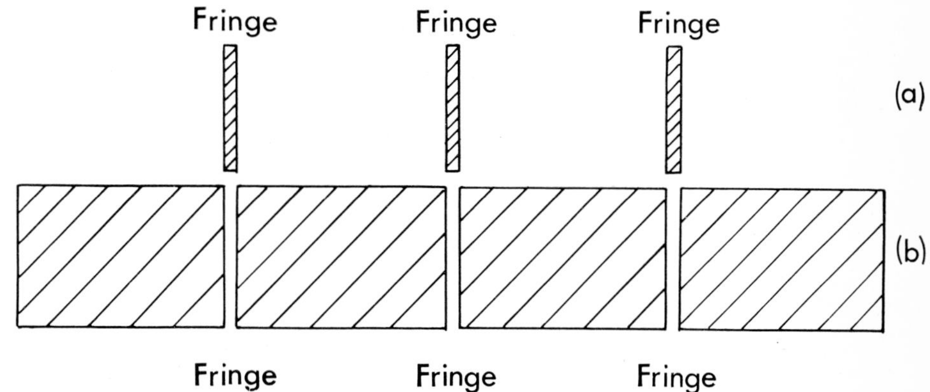

FIGURE 5 Schematic shape of multiple beam fringes (a) in transmission (b) in reflection

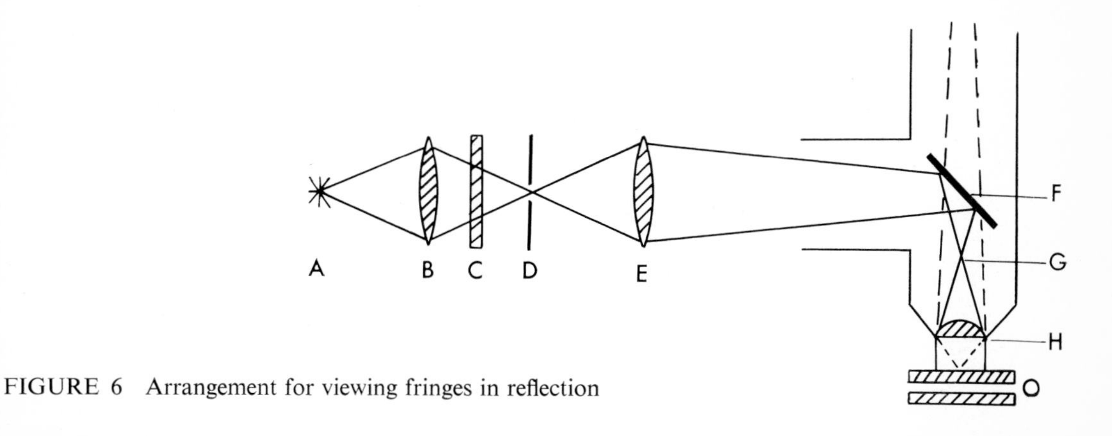

FIGURE 6 Arrangement for viewing fringes in reflection

Chapter IV

Specular Surfaces

Polished diamond

The mechanical polishing of surfaces, whether that of plate glass, silver utensils, surfaces of engine pistons, diamond gem stones, etc., has long had important technological applications. Such polished surfaces are admirable subjects for study by multiple-beam interferometry. Here we shall select one

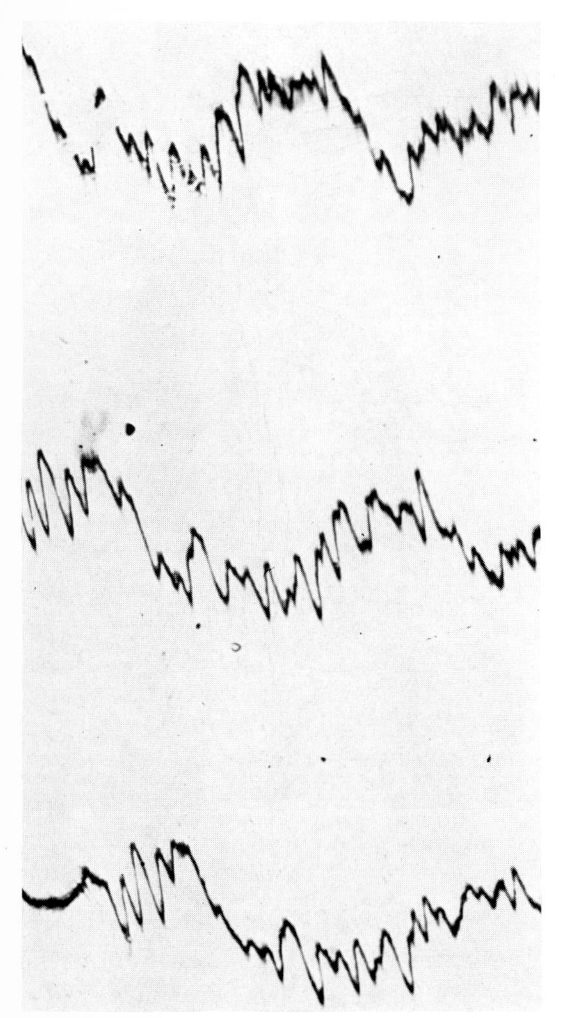

PLATE 4 Interferogram over the face of a commercially well-polished diamond. The zigzag reveals residual polish ruts

PLATE 5 Improved smoothness on a polished diamond, but only achieved after prolonged uneconomic polishing

or two examples which will illustrate potentialities.

It has been known for perhaps six centuries that a brilliant 'adamantine' polish can be secured on the surface of a diamond gem stone. It was formerly believed that this implies that a diamond takes a perfect polish. Interferometry indicates that this is far from the truth. Plate 4 shows a reflection interferogram given by the "table" face of a commercially polished "brilliant" diamond. This is the main face on a gem stone. The fringes zigzag along merrily.

Diamond is a crystal which offers resistance to polish differently in different directions. Some directions are harder than others. Only by exploiting the 'hard' direction of one diamond can another diamond be polished, and then only in its 'softer' directions. (Hence the folk-lore saying 'diamond-cut-diamond'.) Indeed the zigzags in Plate 4 do actually expose the inherent directional hardness variations on the crystal. Prolonged polishing well beyond ordinary commercial practice can much improve the finish, at extra cost, but the secondary zigzag can never be wholly eliminated.

It is significant that the zigzag features extend only over a small fraction of the separation between fringes. (The distance between one pair of fringes is called an 'order' of separation.)

The 'waviness' is more or less one-eighth part of an order, i.e. 1/800,000th inch. This would hardly be seen at all with light interference only exploiting two light beams. It is only the high sharpening of *multiple-beam* interferometry which exposes the fine detail rugged surface structure. Magnification in this picture is about $\times$ 200,000 and resolution is very high, some detail of 1/100th of an order being easily resolvable.

Prolonged (and uneconomic) polishing improves matters to the position shown in Plate 5, which still shows the residual directional hardness effect persisting.

PLATE 6 Step ledges on a mica cleavage surface

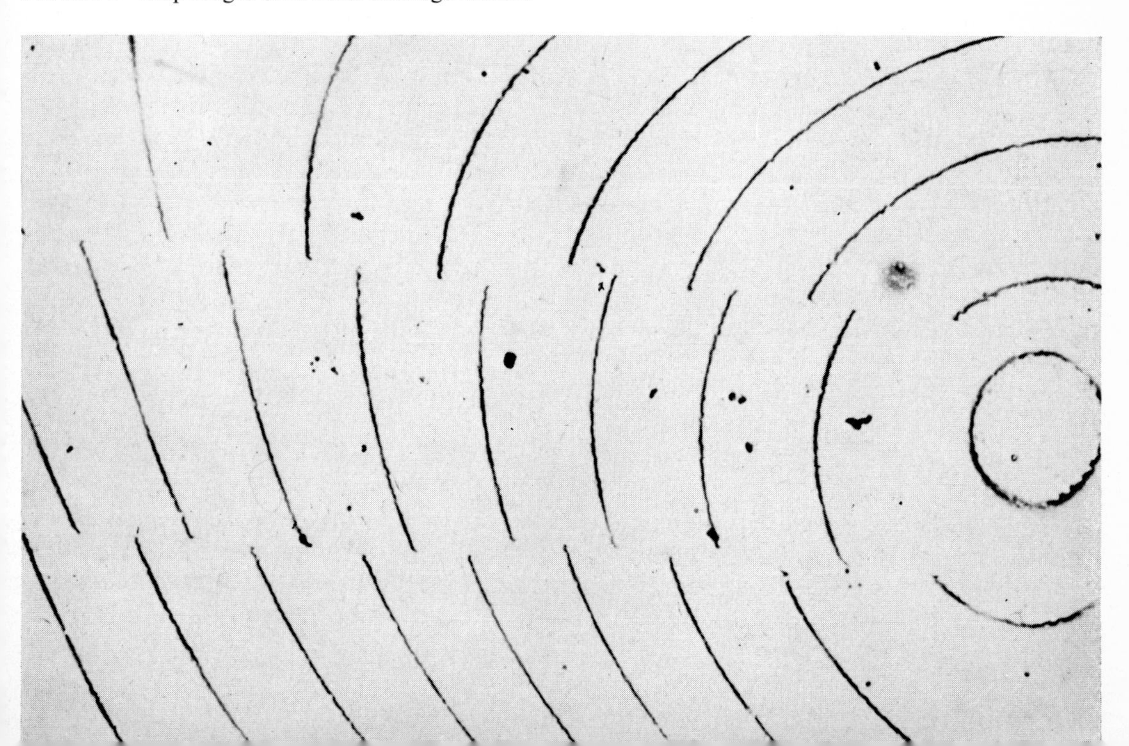

One therefore asks oneself, are there many, or indeed any, natural objects which are really smooth? There are in fact very few, but there is one outstanding and remarkable example; this is the mineral mica.

Mica cleavages

Mica is a crystalline material which is found in sheets, sometimes more than a square foot across, and it has the remarkable property of exhibiting a perfect cleavage. By inserting a fine needle into this solid rock crystal very thin sheets can be coaxed off, in fact, with luck, sheets as thin as 1/25,000th of an inch. So thin a sheet, which is only two light waves thick, is not easy to cleave off but with perseverance such sheets, about a square inch across, can be obtained. Mica is heat resisting and many will be familiar with its use in the windows of anthracite stoves. It has many electrotechnic uses as an insulator of high quality.

Now we have found that the surface of a cleaved sheet of selected mica can show a superb degree of natural specular smoothness, but although very smooth it is never flat. Plate 7 shows, fifty times magnified, a multiple-beam interferogram given by the cleaved surface of a very good-quality piece of mica. There is much to be learnt from this. It looks astonishingly like our Plate 1 contour map of geography, with its hills and dales. The fringes are fine and narrow, showing that a good optical technique was used to obtain them. The fringe smoothness shows up clearly the natural high specular quality of the mica surface, although small significant irregularities can be noticed here and there.

A notable feature which appears on numerous mica cleavage surfaces is the appearance of step line discontinuities, a typical example of which is shown by the fringe pattern of Plate 6. Each line discontinuity is revealing a cleavage step, a kind of 'micro cliff edge'. All crystals are built up from the repetition in space of a small regular atomic grouping called the unit cell, which repeats to build up the final crystal. The distances between the atoms which form the unit cell are of the order of one or two angstrom units or less. The spacings between the planes of atoms which are created through multiplication of the unit cell are called the *lattice spacings*. When a crystal can cleave easily it is because in its atomic groupings there exists a relatively weak binding force across some lattice spacing, enabling a crystal to be parted relatively easily on this plane. The plane is called a *cleavage plane* and of course also repeats. The distance between such repeat planes is called the *cleavage lattice spacing* and this can be found by X-ray methods of crystal analysis.

It has been established by X-ray methods that the cleavage lattice spacing in mica is almost exactly 20 angstrom units. We find indeed that the cleavage steps which interferometry reveals are in fact 20, 40, 60 . . . angstrom units. They are thus small integral multiples of the lattice spacing. So closely is this followed that we have been able to evaluate lattice spacings for several other crystal types which cleave, using the interference fringes.

PLATE 7 Hill and dale interferometric fringe pattern shown by the surface of a cleaved sheet of good-quality mica

Chapter V

Surfaces of Natural Diamond

Diamond crystals

Even in Biblical times it was recognised that diamond was the hardest material then known to mankind, and despite modern technology, remarkably, diamond still remains the hardest material known to man. This makes it an object of curious interest to the scientist and at the same time an object of strategic importance in engineering technology. As a material for trueing and machining other very hard material, there is nothing which can match up to diamond in this respect. This is partly due to its hardness, partly to its good thermal conductivity and partly to its low thermal expansion.

Because diamond is also very resistant chemically, it can be expected that, when removed from the earth, it is not likely to have suffered either from abrasion or from destructive chemical attack. Thus its surface may well be expected to retain its original growth characteristics. It so happens that the mode of origin of diamond growth deep down in the earth is still a considerable mystery but some light is being shed through interferometric studies on its microstructure, although controversy still goes on as to how diamond really grows.

Diamond is one form of crystallised carbon, graphite being the other. Since diamond was first synthetically made in 1955, today many millions of small diamonds are being commercially synthesized from graphite, in amount equal to several tons in weight, but it is quite certain that natural diamond and synthetic diamond grow in very different fashions. Natural diamond is found in four main kinds of shapes, namely: (1) *Octahedra*: eight-sided figures, something like two Egyptian pyramids stuck base to base; (2) *Dodecahedra*: twelve-sided crystals, each face side being rhomb-shaped and usually curved; (3) *Cubes*: cubic in shape, but with very rough surfaces; (4) *Triangular plates*: really a type of modified octahedron, called a twin.

On very rare occasions a hexagonal plate appears. A group of diamonds from the writer's own personal collection is shown a little enlarged in Plate 8. The unusual hexagon plate in the centre is over one centimetre across. In the picture can be seen all the types mentioned above.

The eight-sided octahedra have attracted a good deal of scientific attention, in particular because they occur very frequently and also because of some remarkable features which often appear on their eight triangular faces. Argument about these features still goes on, but whatever their cause they are notable objects.

PLATE 8 Varieties of shapes of diamond crystals

Diamond trigons

On the triangular octahedral faces there often appear small triangular depressions which are called *trigons*. Sometimes there are relatively only a few as in Plate 9; sometimes there are vast numbers on one face, at times over a million. When interferograms of these trigons are taken, astonishing micro-topographies emerge. Plate 10 at a magnification $\times 100$ shows fringes on a region on one surface and we see here two kinds of trigon, those which are relatively large, with successive concentric fringes, and those which are small, appearing as fairly uniform triangular light patches. The larger ones are pyramidal hollows, going down to a point. Remembering that each fringe indicates a depth change of

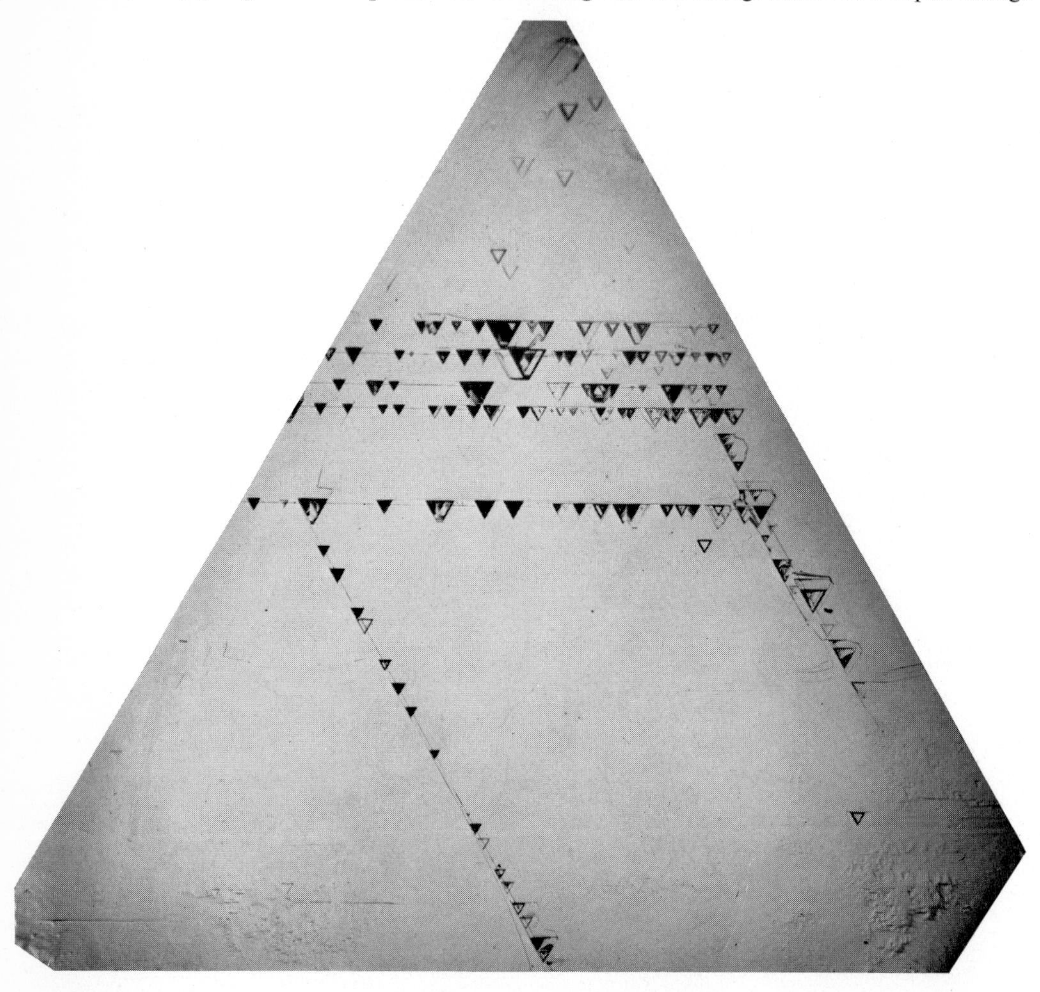

PLATE 9 An enlarged triangular octahedron face of a diamond, showing numerous small triangular pits called 'trigons'

1/100,000th inch, we see that the centre trigon has a depth of say 1/25,000th inch or so. Yet this drop in height takes place in a distance of only about 1/250th inch when magnification is allowed for, so the slope of the side is very shallow indeed, i.e. about 1/1000 radian, say 1/20th of a degree, roughly. The uniform tint trigons are shallow flat-bottom pits, many of which are (it can be shown) a mere 1/500th part of a light wave in depth, say 1/25,000,000th inch!

A bewildering variety of trigon structures has been found. One such picture is Plate 11, which shows a profusion of trigons of various depths. The trigons are all strictly oriented in the same direction. Such an interferogram contains on it a wealth of information about the shapes and depths of the trigons.

PLATE 10 Fringes, at magnification $\times 100$, over trigons

PLATE 11 The complex interference pattern on a diamond face which has on it a profusion of trigons

Etched diamond

Although diamond chemically resists the fiercest of acids and alkalis easily, even when they are very hot, yet because it is pure carbon it can be attacked by hot oxygen (to burn away to carbon dioxide). It is only slowly attacked by the liberated oxygen if it be heated to about 600°C in potassium nitrate.

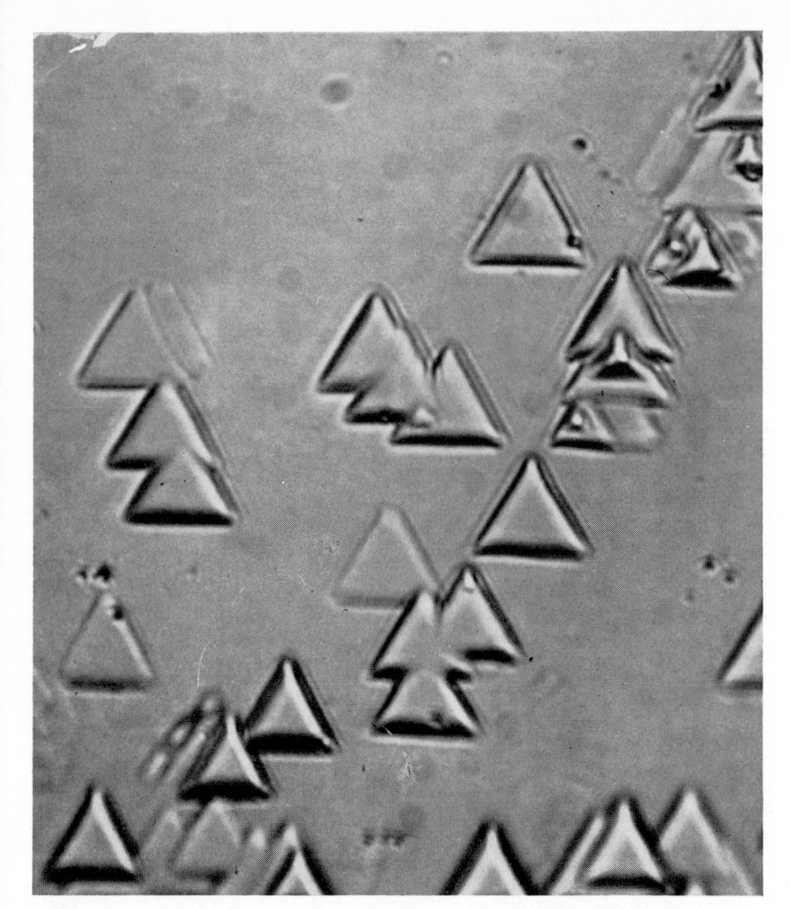

PLATE 12 Small etch pits artificially produced on a diamond face by hot oxygen

The attack first takes the form of producing small triangular hollows called etch pits which look at first very like flat-bottom trigons but are *oppositely oriented* to the trigons. For if we look at Plate 9 we shall see that the natural trigons lie such that in each the triangular apex points to the side edge of the triangular face. The potassium nitrate etch pits on the contrary point the other way, they are oriented parallel to the crystal face. Plate 12 shows some small etch pits magnified.

When the rate of etching is increased these pits grow bigger and deeper and begin to round off. An interferogram can reveal their microtopographic shape. Plate 13 ($\times 250$) shows the quite complex patterning resulting from a prolonged etch. Such fringes often show a good deal of subtle detail enabling calculations of shapes and depths to be derived. Especially notable is the way in which pits show discontinuities internally and very often the fringes indicate how the deep pits eat into each other.

When the process of etching is continued this mechanism of eating into each other advances to a stage where the depth of etch becomes too big to be examined by interferometry. Already this stage is nearly being reached in Plate 13. For here the central pit, looking so like a thumbprint, is some 70 fringes deep, which is about 1/60 mm or so. At perhaps double this depth the fringes would be almost blurring into one another. As etch advances still further the diamond acquires what to the naked eye looks like a frosted appearance, yet under the microscope this is seen to be a remarkable minute 'block' pattern. Plate 14 (at $\times 2000$) shows one example. It has been nicknamed in our laboratory the 'Diamond Manhattan Skyline'.

It so happens that this kind of photograph, which was secured with oblique illumination, is subject to a most curious optical illusion worthy of mention. It will certainly be agreed that Plate 14

PLATE 13 Interferogram given by a deep pit etched on a diamond

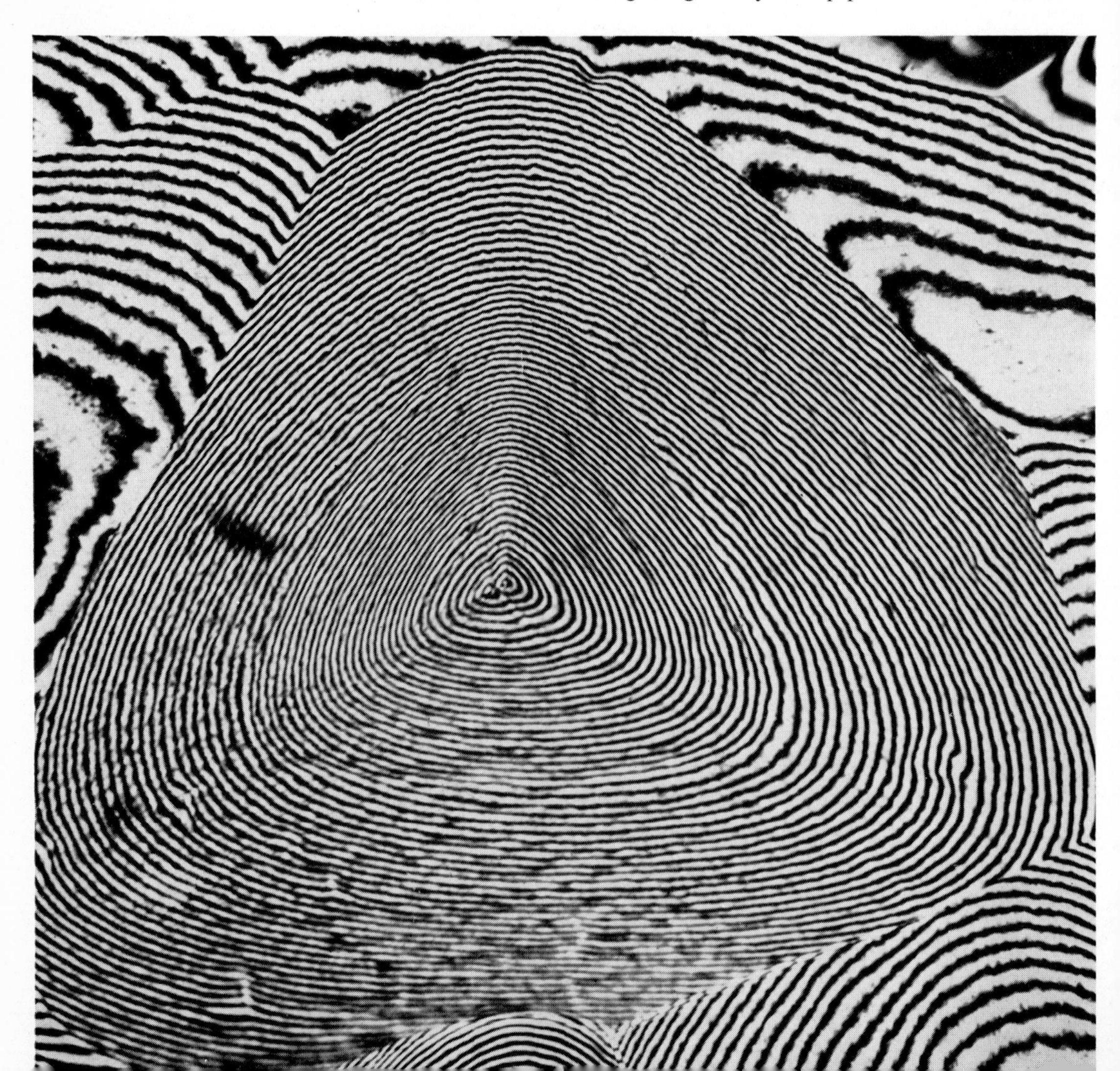

gives the firm impression of consisting of solid blocks standing up proud, piled on top of each other. If the reader slowly turns the page upside down he will have the strange impression that he is looking at hollow re-entrant caverns instead of blocks sticking out. This unusual optical illusion occurs in other types of photograph for instance in aerial photography.

PLATE 14 Block pattern resulting from extensive etch on a diamond face. Magnification × 2000

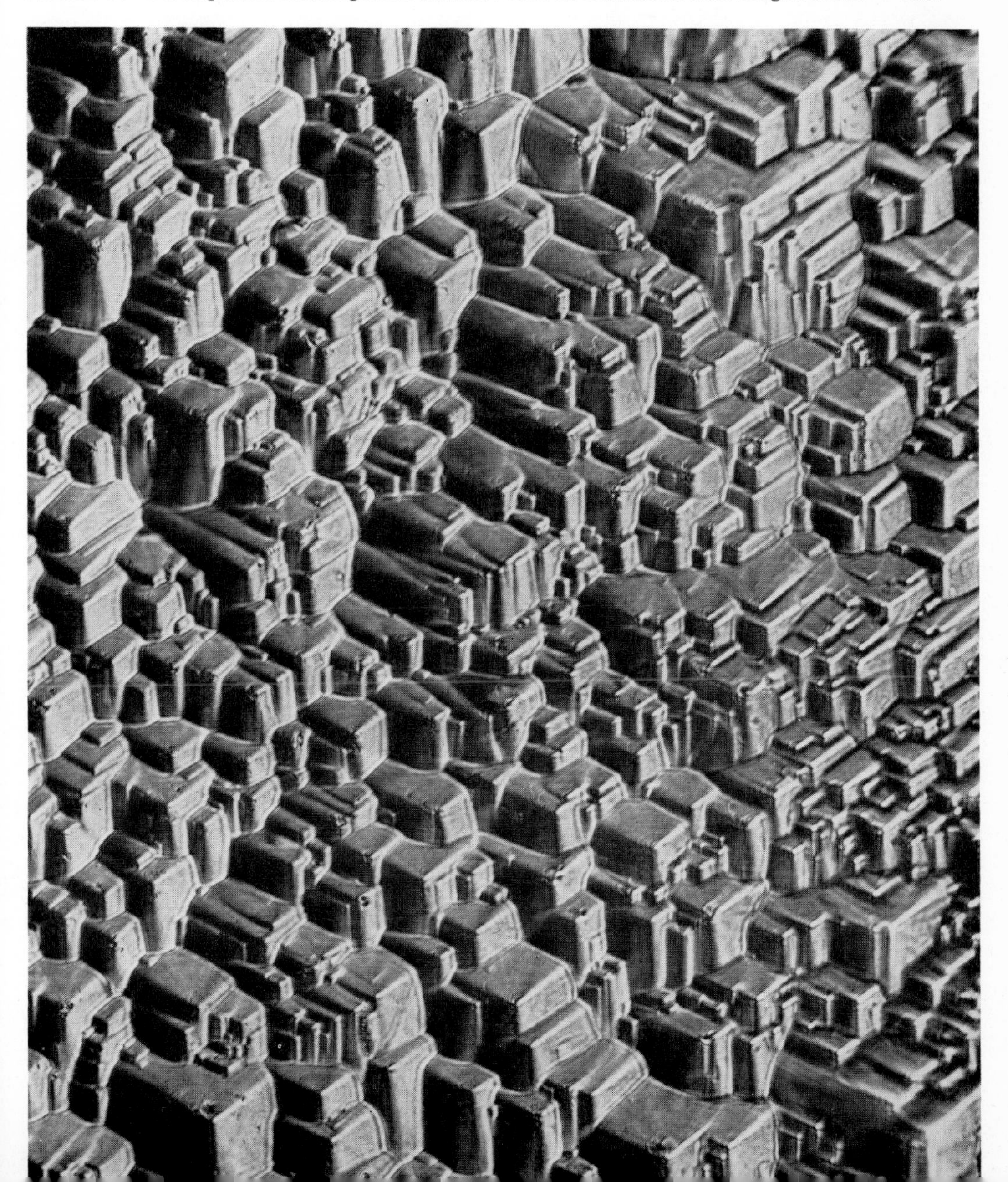

Stratigraphy

We also discovered during our studies of diamond that a very carefully controlled etch has the remarkable power of revealing a complete *stratigraphical growth* history of a diamond. There are good reasons for believing that an octahedron diamond grows by successive layers deposited on the triangular octahedral faces. If we truncate such a diamond as in Fig. 7 then the sectioned plane, which is called cubic, cuts across all the growth sheets. Now when such sheets are being laid down they do not always key perfectly. Furthermore, there is evidence that some sheets appear to occlude nitrogen,

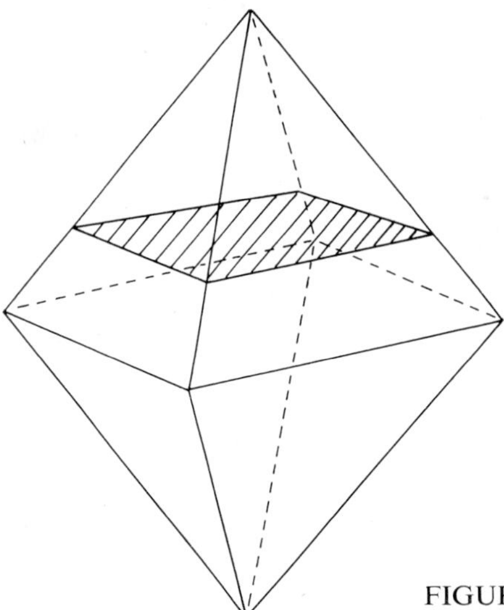

FIGURE 7 Diamond sectioned on a cubic plane

which seems to concentrate locally in very minute platelet regions. When such a truncated surface is subjected to a very critically controlled slow etch, the oxygen preferentially first attacks either the weak boundaries between misfit sheets, or the regions which have the chemical impurity (e.g. nitrogen in platelet form). The consequence is that careful etching of the cube plane succeeds in revealing the growth history of the sectioned diamond. Plate 15 is a notable example. This shows the etch pit pattern across a face about 5 mm square. The surface is obliquely illuminated such that light is scattered back strongly to the eye from regions roughened with the minute etch pits. The dark (non-scattering) regions are those which have only very few etch pits and these clearly are regions of good quality crystal, or at least regions which are free from localised nitrogen platelets. The bright areas show the impure layers. A picture like this offers much valuable information about the whole growth history of the crystal. Such a stratigraphy is normally invisible and goes unsuspected. It is only exposed through the trick of controlled etching. Very careful control is essential for, if etching proceeds just a little too far, then the whole surface etch-roughens and the phenomena sought for are completely obliterated, never to be recovered, unless a new section is cut.

We have sectioned large numbers of diamonds and such stratigraphical patterns are found to be quite normal and common.

PLATE 15 Stratigraphical pattern appearing on sectioned cubic plane of a diamond, after subjection to light etch

3

It is of interest to note that, despite the considerable sensitivity claimed here for multiple-beam interferometry, the fact remains that the polished face of the crystal of Plate 15, prior to etching, showed no noticeable features other than polish marks. Interferometry did not reveal the stratigraphy, for indeed the inherent natural stratigraphy does *not* lead to surface height alterations on polishing. This only shows how remarkably sensitive etch is to slight dislocations and impurities, indeed probably sensitive to features which are less than a single crystal lattice spacing, i.e. to dislocation effects of perhaps the order of one angstrom, if not less.

Crossed fringes

We now draw attention to a special combination of interference fringe-systems which leads to a truly three-dimensional picture of a surface. This is illustrated on one of the rare hexagonal plate crystals of the kind shown at the centre of Plate 8. If we look again at Plate 10 it becomes clear that the interference fringes are not only giving contour lines of deep trigons, but are also revealing very shallow trigons through a change of intensity. This can be described as producing *interference contrast*. What was before shallow and invisible is now brought out into intense contrast by multiple-beam interference. This kind of contrast pattern arises only when the reference flat is brought nearly parallel to the object. Then, the fringes are widely broadened by the increased dispersion and small changes in height lead to very marked alterations in intensity.

Using this technique we then exploit what we call 'crossed fringes'. First a picture of the crystal is taken in the parallel setting, i.e. an interference contrast picture. Without removing the plate from the camera the two surfaces crystal and flat are then tilted to a wedge and thus sharp narrow multiple-beam *line* fringes are superposed on the former background of tint variation interference contrast fringes. Then, again, without moving the plate a third exposure is superposed, this time with the surfaces tilted to form a wedge more or less in the direction *perpendicular* to the former wedge.

The developed plate, taken at magnification $\times 50$, with its three sets of fringes, i.e. background contrast plus two crossed grid-like line fringes, is shown in Plate 16. This yields a three-dimensional picture of the whole microtopography. Growth sheets, shelves, triangular depressions curved hillocks and so on can be seen. All become amenable to exact interpretation and to precise measurement. Indeed a comprehensive insight into the shape and topography of this surface is easily obtained. This specimen does happen in fact to have a relatively simple surface compared with some.

The picture has been taken, intentionally, with three separate light waves, one green and two yellow lines in the mercury spectrum. Between the single green fringes appear the yellow doubled pair fringes. The fringe order is from single to single (or pair to pair). On pursuing along a line fringe it is seen how the height mounts or falls over stepped ridges, and how the background fringes give the clue to the meaning of the various wriggles, steps and jumps of the line fringes.

Some multiple-beam interferometry has been most successful in revealing hitherto unsuspected microtopographies on many other minerals and crystals as well as on diamond .We shall devote the next chapter to an indication of the kind of information already obtained with the important mineral quartz. The studies described concern not only natural features on quartz crystals, but also some studies on quartz crystals when used to control electrical oscillations.

PLATE 16 'Crossed' fringes on a diamond face. Three different sets of fringes are here superposed

Chapter VI

Quartz Crystals

Crystal faces

Beautiful quartz crystals (rock crystal) are often found in nature and they can vary in size from small specimens, like grains of sand, to fine massive blocks over two feet in height! Quartz is used for a great variety of scientific purposes, so much so that nowadays, after much effort, it is possible to grow fine quality, fairly large, synthetic quartz crystals which adequately provide for many of the needs of technology. For long the natural quartz crystals have excited much attention because of the beautiful shiny smooth nature of their end faces. The crystals tend to grow out of a matrix rather in the form of long six-sided pyramids as in Fig. 8, having usually ridge markings, called *striations*, on the long rectangular faces. These, as shown, are capped often with shiny smooth triangular faces.

There are two kinds of alternating terminal faces on the ends of the hexagonal pyramid, the large, called *R*, the small *r*. It is these end faces which are usually so smooth and so shiny that an interferometric examination is invited and when undertaken it reveals quite notable characters on selected crystal faces.

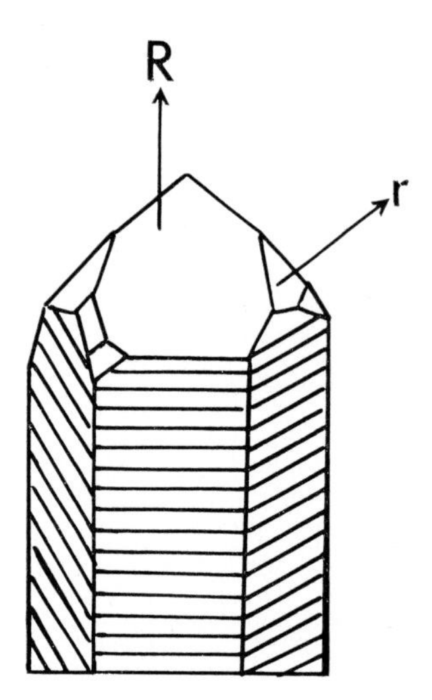

FIGURE 8 Schematic outline of a quartz crystal

Although appearing very smooth to the eye, the *R* faces always, interferometrically, show up a shallow rounded pyramidal structure. Plate 17 shows, at $\times 20$, a fringe pattern from such a face. The smoothness of the fringes shows that the surface has a good specular quality. There are here three clear triangular pyramidal hillocks, tetrahedra, with slightly rounded sides and corners, of height, perhaps some four fringes. Not all such faces are so smooth as this by any means. Sometimes slight shallow striations appear and Plate 18 ($\times 40$) shows fringes running over a striated surface. Clearly

PLATE 17 Fringe pattern from a smooth, end (R) face on a quartz crystal

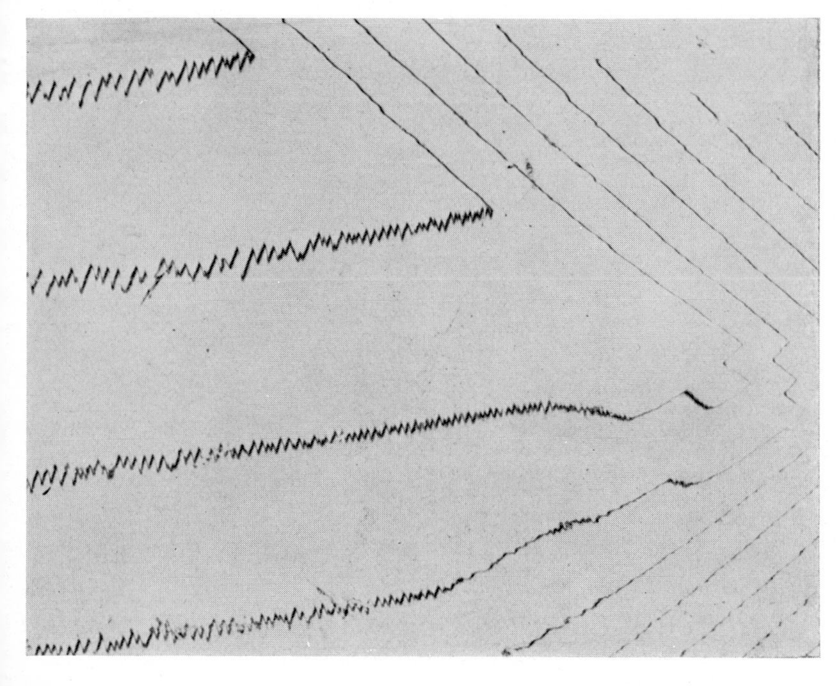

PLATE 18 Fringes showing very shallow striated markings on the face of a quartz crystal

the striations are quite shallow sharp ridges over which the fringes zigzag, the depths of these ruts being mostly in the region of 1/20th to 1/10th of a fringe separation, i.e. a mere 1/2,000,000th to 1/1,000,000th inch. A picture like this does strikingly reveal the power and sensitivity of the multiple-beam fringe technique for exposing the micro-contour details of surfaces.

Such striations are growth sheets and under the illumination of a good microscope such sheets can be faintly seen. Yet although such pictures convey the information that the striated sheets are indeed bunching together, they are merely qualitative pictures. To obtain a real numerical interpretation of bunching of such growth sheets we need to turn to interferometry; Plate 19 ($\times 60$) shows in striking fashion the way in which a succession of growth sheets on such a face bunch together. The exact numerical value of the step heights can readily be assessed from the interferogram.

The small *r* faces are also found usually to consist of well defined triangular hillocks, but superposed on these is often much fine growth-sheet layer striation, on an extremely shallow scale. Plates 20 and 21 ($\times 60$) show two such microtopographies. Close inspection reveals that the fringes are running over extremely shallow striations, many of depths within the limit of resolution of the already highly sensitive interference method. They must be only a few atoms high. There is much information on such pictures, but a further discussion would lead us into special crystal growth problems.

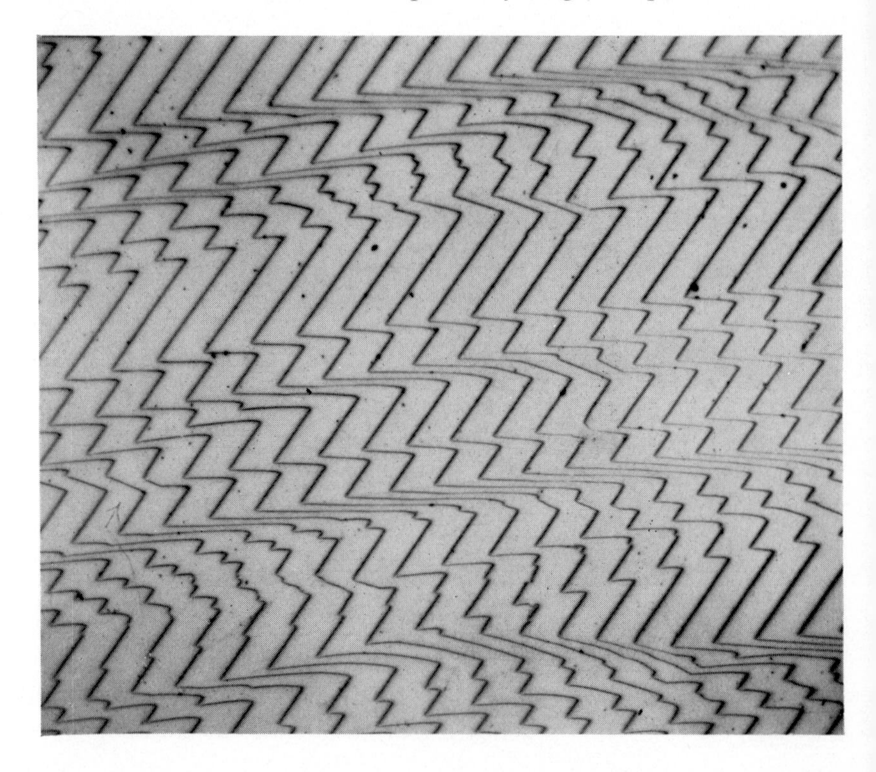

PLATE 19 Bunching of growth sheets on the face of a quartz crystal, revealed by an interferogram

PLATE 20 (above right) Interferogram showing topography on a smooth, end (*r*) face of a quartz crystal

PLATE 21 (below right) Complex pattern showing also shallow striated growth features on a quartz (*r*) face

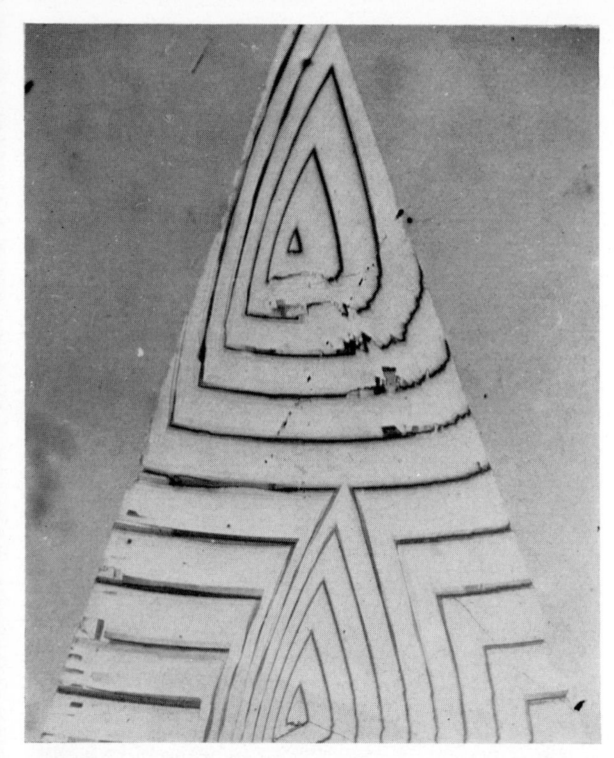

Quartz crystal oscillators

Quartz crystals are *piezoelectric*, that is to say, they have the property of expanding (or contracting) when placed in a suitably directed electric field. If a slab of quartz is placed between two electrodes to which an alternating potential is applied the quartz is forced into mechanical oscillation. According to its dimensions, and the particular way in which the slab has been sectioned out of the parent crystal, each quartz slab has its own natural modes, with its own frequencies of vibration. It can be induced to vibrate in various ways due either to flexure, to shear or to torsion. Various modes of vibration are possible according to shape. Furthermore some of these different types of oscillation can interact with each other. Just as a crystal can be forced into different frequencies, so conversely a vibrating crystal can impose its own frequency (when properly arranged) on to an oscillating valve circuit. Since the frequency of vibration of quartz is remarkably steady and on a correctly cut crystal is hardly affected by temperature, pressure or other disturbances, the quartz crystal oscillator becomes a powerful tool for controlling and *keeping highly constant* the frequency of a radio system. Quartz frequency control is so accurate that quartz 'clocks' have been constructed which are better timekeepers even than the earth in its spin and rotation round the sun! Indeed, by comparing independent quartz crystal clocks which agree with each other it has been found that slight irregularities in the earth's movements have been detected. In radio technology quartz crystal frequency control is now of fundamental importance.

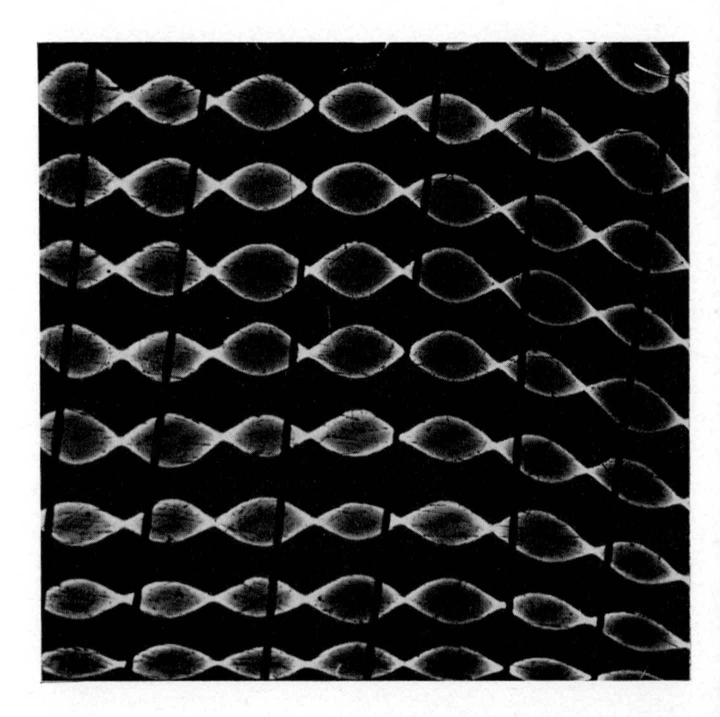

PLATE 22 Fringe pattern given by a quartz plate set electrically into high-frequency oscillation. Nodes and anti-nodes are revealed interferometrically

Now it so happens that we have been able to use multiple-beam interferometry as a method for studying both the *distribution of any nodes and anti-nodes and also amplitudes* of vibration for an oscillating quartz crystal. As a rule, quartz crystal vibrators are cut in special shapes, e.g. rectangular blocks, plates, flat circular discs, or ring shaped annuli.

The frequencies of the crystals may conveniently range from say 10,000 to 1,000,000 vibrations per second. At first sight it appears formidably difficult to try to use interferometry to examine surfaces of such a rapidly vibrating object. Yet, in fact, the interference technique is found to be quite simple. One face of the polished quartz crystal is silvered and this rests on a silvered optical reference flat so that *straight-line* interference fringes are formed between crystal and flat. A number of the straight-line fringes are arranged to cross the field of view, say ten or more when the crystal is at rest. Then the alternating electric field is applied across the crystal to make it oscillate. In general there is some up-down movement relative to the optical flat. Clearly at the nodes the distance between crystal face and reference flat remains unchanged so nothing happens there to the fringes. But at anti-nodes the distance between crystal and flat varies in a sinusoidal way. As a result the fringes broaden out. Thus not only is a nodal pattern revealed, showing where and how the vibration is taking place, but in addition, no matter how quickly or how slowly the crystal vibrates, the stabilised local fringe broadening *gives an exact measure of the local amplitude*. It is thus possible to study relationships between current and amplitude, damping effects and so on.

We start with a rectangular quartz plate. When at rest this was crossed by a number of straight line fringes. When set into oscillation the plate flexes in a regular way such as to show a succession of linear nodes. From the fringe broadening the local amplitude can be exactly derived. Plate 22 shows the pattern given at a particular degree of energising the plate. Inspection shows local alterations in amplitude, which had not been anticipated. It is to be remembered that before oscillating the fringes were simply sharp, narrow lines. The dark lines crossing the picture are merely obscurations from a number of wires acting as an electrode.

Vibrating discs

The oscillation interferograms given by vibrating quartz circular discs are of particular interest. When the disc is placed in a circuit and set into oscillation, by regularly varying the capacity in the oscillator circuit it is possible to excite the disc into a whole, controllable, set of different kinds of vibration, i.e. different modes. Each mode is linked to a particular frequency and can be reproduced at will by adjusting the frequency. At any mode the crystal oscillates in a steady fashion in its own special manner. Interferometry reveals the mode of oscillation in a striking way. Thus, for example, Plate 23 shows one particular thin quartz disc oscillating at a selected frequency. It can be seen that the centre is at rest (nodal) and this is surrounded by three concentric ring nodal regions. Superposed on these are diametrical linear nodes crossing more or less at 60°. Yet this is only a part of the information available. For by the widening of the fringes in any part, the exact amplitude of local traverse of vibration is readily obtained. It is also of interest optically that the nature of the fringes is such that considerable overlap of orders is possible without serious confusion. The decrease in amplitude on going out radially from the first anti-nodal region, towards the circumference, is notable.

On increasing the frequency, a vibration pattern like that in Plate 24 is secured. It is interesting to hold this at arm's length, or further, when it gives a vivid impression of the distribution of nodes and anti-nodes.

Considering that the crystals can be vibrating at many tens or hundreds of thousands of cycles per second, it is remarkable that such stable patterns appear, since a photograph of the fringes may need an exposure of, say, ten minutes.

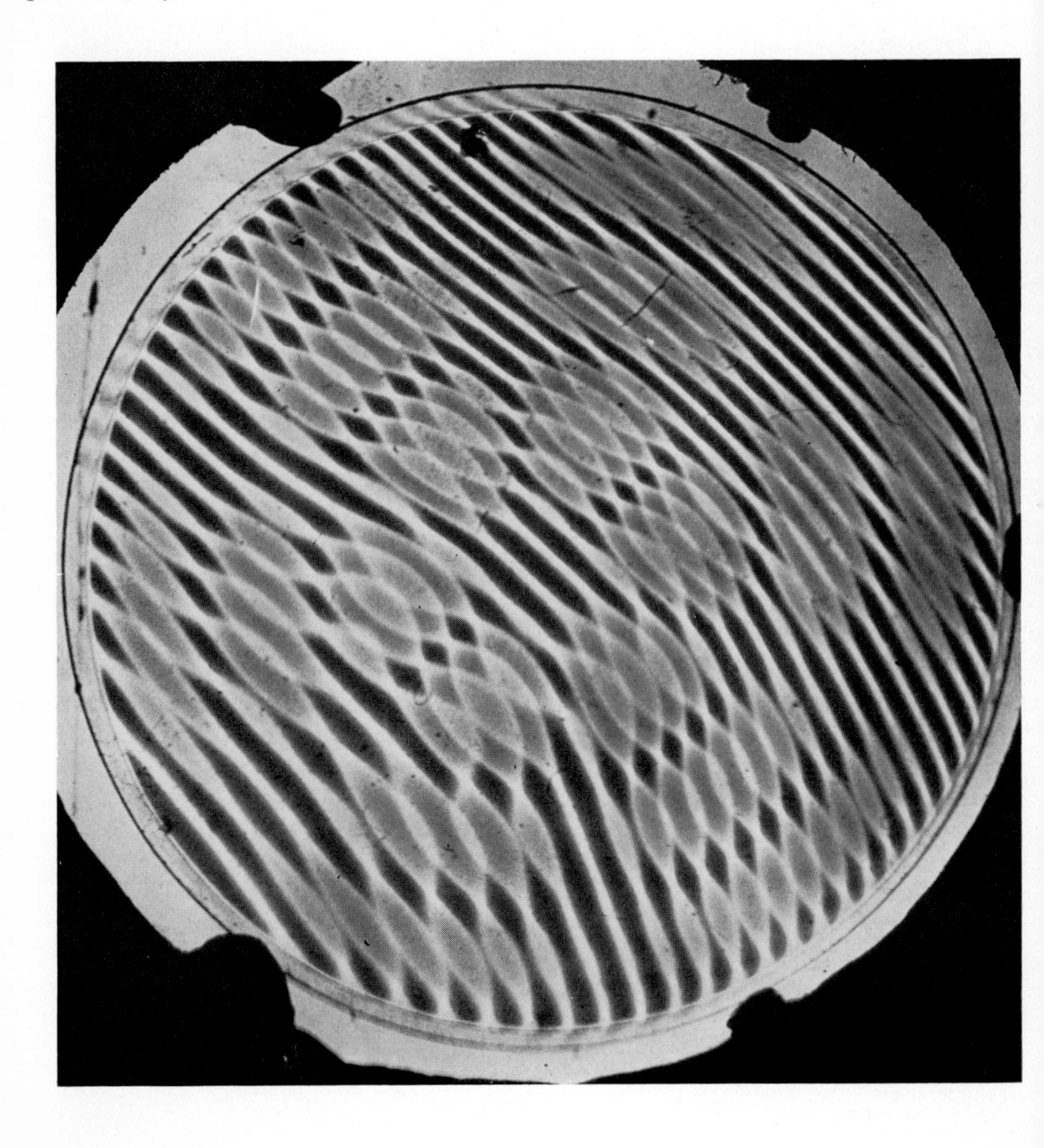

PLATE 24 (above) Fringe pattern from a quartz disc vibrating at an increased frequency and showing higher overtones

PLATE 23 (left) Interferometric pattern from a vibrating quartz disc

Quartz clock

As a further example we illustrate in Plate 25 a fringe pattern given by one of the high precision quartz clock annuli designed by Dr L. Essen of the National Physical Laboratory. Such an annulus has especially good temperature characteristics. It is seen here oscillating in such a fashion as to give twelve radial nodes. A very big amplitude has been forced on the crystal, yet despite the multiple overlap of fringes each can be identified. The vibration fringes, like those in Plate 23, reveal a most interesting optical feature. Because the crystal is vibrating in a sinusoidal fashion, each section comes to rest at the extremes of its amplitude and then swings through with maximum speed at the centre of its vibration pattern. Thus the time occupied at the end of the vibration is much more than the time at the centre. Consequently the photographic exposure time at the end of each swing is greater than anywhere else in the swing. So the fringes are all bounded by a photographically intense enveloping boundary. As a consequence of this, not only is it easy to follow through, without any blurring or confusion, fringes which may even have overlapped *several* orders of interference, but also, because each oscillation fringe ends in a *hard sharp boundary*, the excursion of vibration can be measured with very high accuracy indeed. Just at the very edge, where high accuracy is most desirable, there is a finely delineated *hard line*, and not as might at first have been expected, a fuzzy fade-out. Plate 25 is printed in reverse to emphasise the point just made.

In the section which follows we shall describe further novel interferometric observations on crystal surfaces.

PLATE 25 Fringes given by an oscillating precision quartz clock. Overlapping fringes

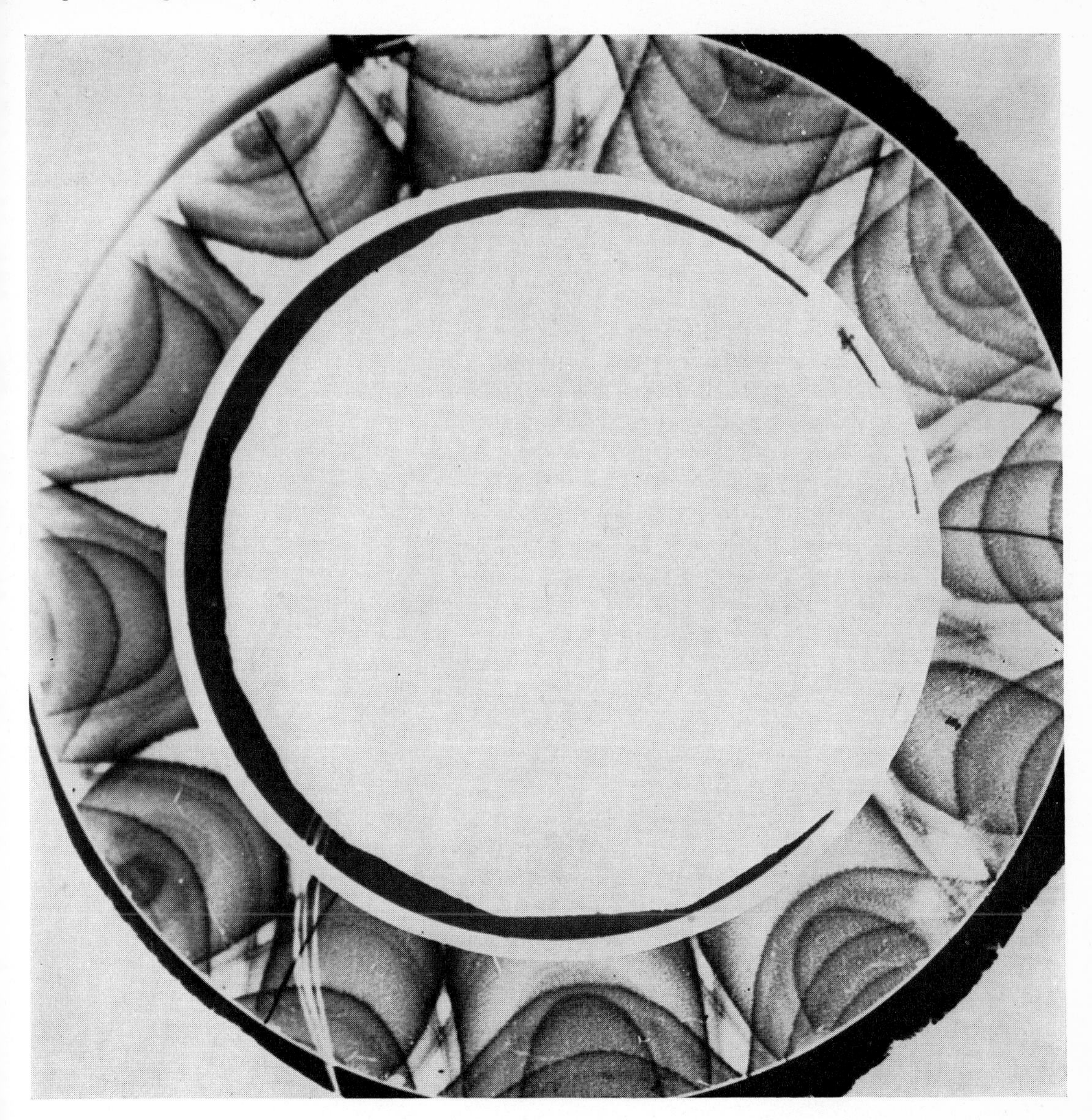

Chapter VII

Spirals on Silicon Carbide

Spiral growth

It has been known for quite some time that a limited number of crystals show on their surfaces peculiar regular spiral formations which are due to a growth mechanism. There can exist on some crystals what is called a *screw dislocation*, a misfit in the array of atoms, which persists through the crystal and emerges on the surface as a linear array of misfitting atoms, originating usually from one point. Theory shows that growth is favoured along a line anchored at this point, as a result of which a growth radius sweeps round in the form of a spiral ramp, mounting slightly higher at each turn. There were very good reasons for believing that the step height of such a ramp could be a small integral multiple of the crystal lattice, indeed it might be just one lattice high, i.e. a few angstrom units. Furthermore theories showed that the growth rate of such a spiral could depend on the crystal directions when there was slow rate of growth (at low vapour pressure if the crystal grew from the vapour or at low concentrations if it grew from a solution). When the growth rate was high, theory indicated that there would be a tendency for the growth to overcome preferred crystal directions so that a circular spiral could result instead of a linear crystallographically oriented spiral.

Although in fact spirals appear on many crystals, it is not usually found that such spirals are regular in shape. For this to happen the condition of both correct screw dislocation and correct concentration is needed. Indeed on many large crystals there may be a single spiral over say one corner, whilst the rest of the crystal has obviously grown otherwise. Complete spirally grown crystals have been found on a tiny micro-scale in various paraffin-type crystals, when examined under high magnification with the electronmicroscope, but as a general rule spirals are hard to find on most common minerals or on other well-formed crystals, and often they change their pattern whilst growing.

An outstandingly frequent exhibitor of spirals however is silicon carbide. This material, the content of Carborundum sharpening stones, is grown in an electric furnace at high temperature and deposits through a vapour reaction of silicon with carbon. On this material we have found many notable spirals and as it is a crystal often with an inherently smooth specular finish, it lends itself to precision multiple-beam interferometry. By this means many theoretical predictions in the theory of spiral growth have been confirmed.

Silicon carbide crystals

Silicon carbide has a hexagonal symmetry and we might expect, if the crystal symmetry imposes itself, that hexagon-like spirals will appear. This is indeed so, but for a first example let us look at Plate 26. This is a very rare example on silicon carbide of an almost perfect circular type of spiral. We cannot tell, looking at it like this, whether we are viewing a hill or a hollow (this is not a foolish proposition, for in some instances, a form of etching can lead to hollow spirals), and this is one reason for applying interferometry.

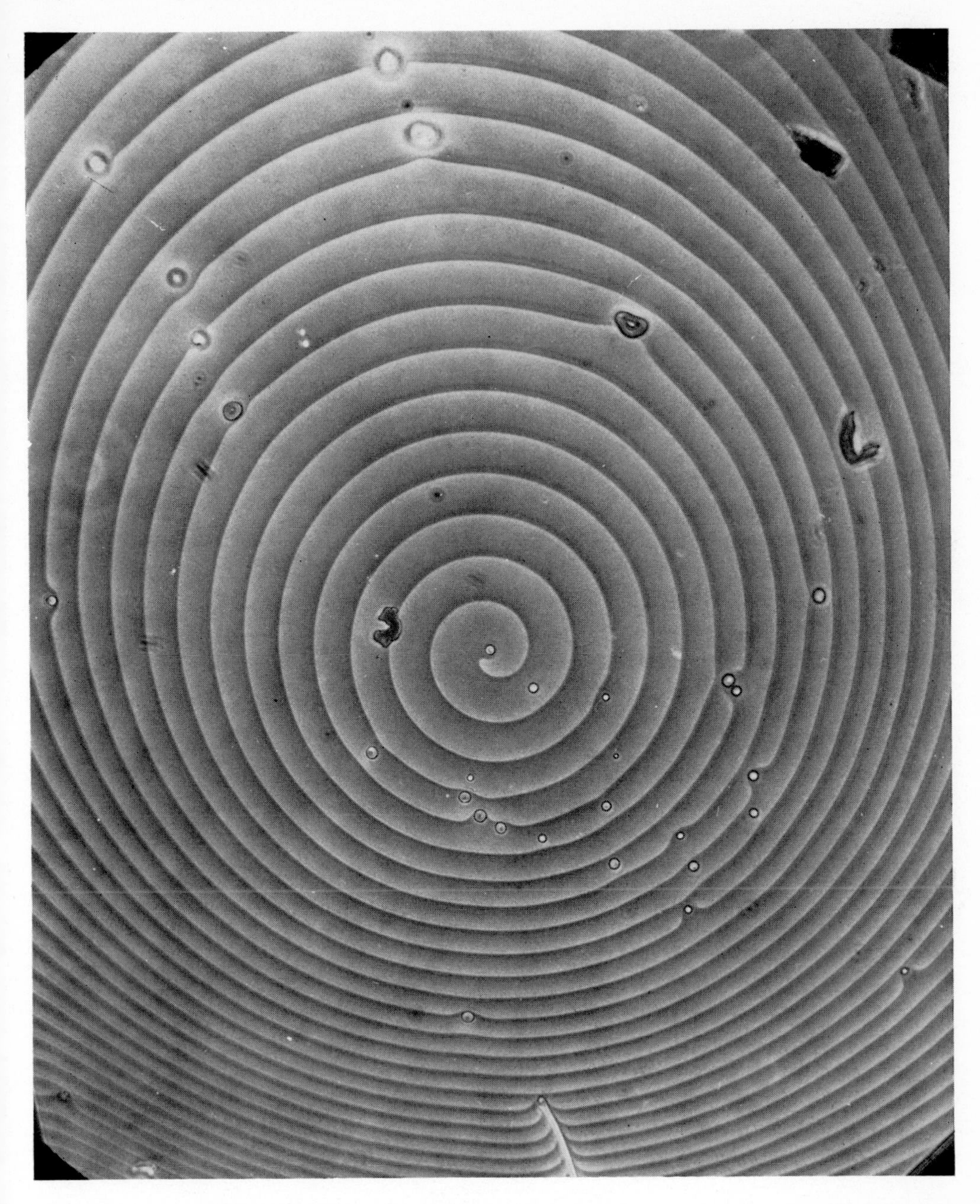

PLATE 26 Circular growth spiral on a crystal of silicon carbide

A much more common type of spiral is that shown in Plate 27 (× 500). Here the dislocation centre spot is just visible. The heart of the spiral is almost circular, but very quickly as it winds round it takes on the anticipated hexagonal character of silicon carbide. This implies that concentration differs at beginning and end of growth. When fringes are run across this quite small spiral we get the interfero-

PLATE 27 Growth spiral on a crystal of silicon carbide. The centre region is circular, then this becomes progressively hexagonal as the spiral winds round

gram of Plate 28. The fringes are so placed that one fringe runs over the peak of the spiral. Because of high magnification difficulties the fringes are broadened slightly, nevertheless their definition is still admirable and this enables a great deal to be deduced. We see that the spiral is indeed a spiral ramp—like a staircase. The steps are all exactly equal in height, except where some obvious defect

PLATE 28 Interference fringes over the spiral shown in Plate 27

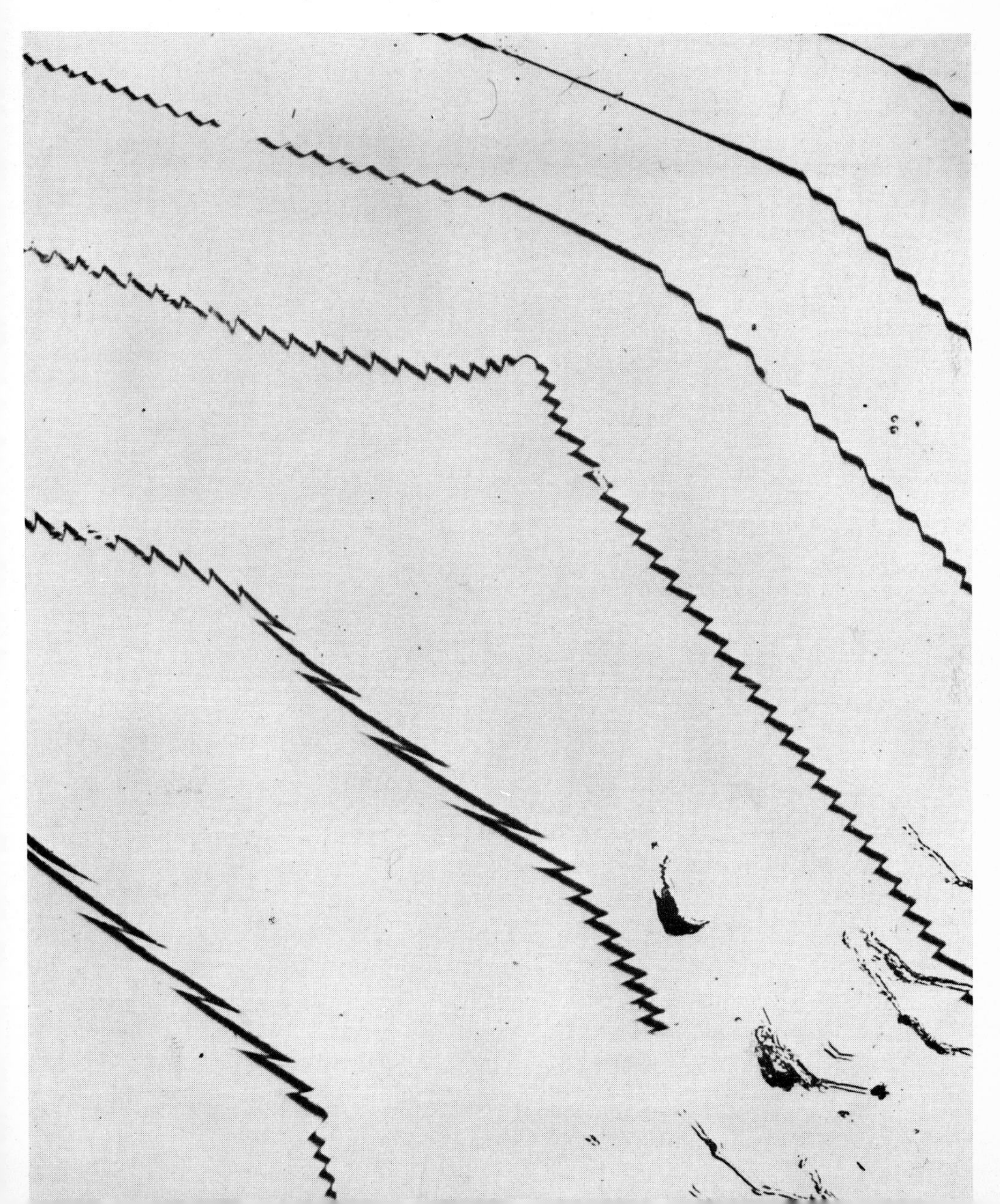

in the surface exists. The smooth flatness between steps is notable. Fringes are roughly 5 cm apart on the original of this picture so that magnification in the up-down direction is about $\times$200,000. The measured step height on this spiral is 165 Å (i.e. about 1/1,500,000 inch) and of course a far smaller step could be measured. There are good grounds for believing that the particular lattice spacing on this particular crystal of silicon carbide is 15 Å, so that the spiral *is a mere 11 crystal lattices in height*. Both much smaller and much larger step heights have been observed on silicon carbide crystals, but all seem to be, as near as can be seen, multiples of the lattice spacing.

Not all spirals are as simple as this one but, if resolvable, there is great perfection in the nature of the spiral ramp. A more complex system with subsidiary hillocks is shown in Plate 29; yet even this is

PLATE 29 Interferogram shown by a complex spiral on silicon carbide

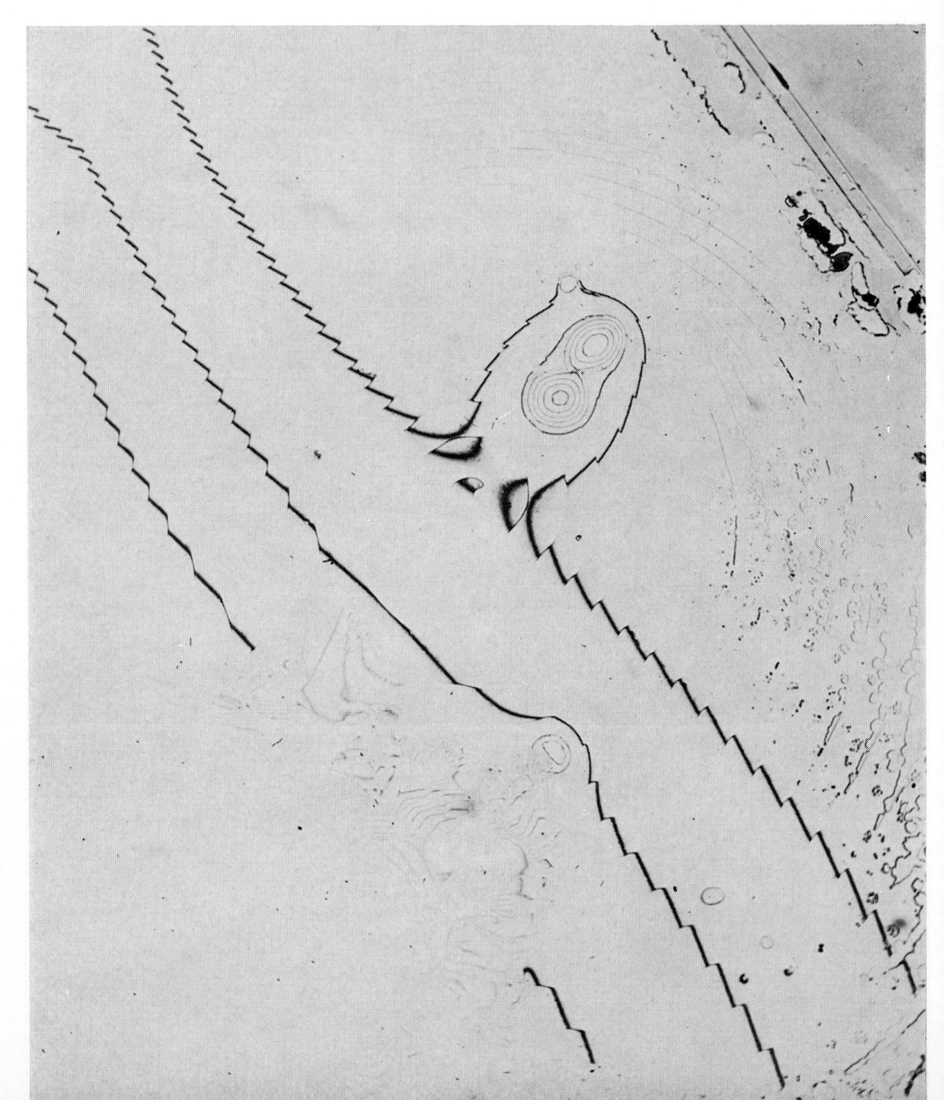

relatively simple. At times an extremely large number of spirals can emerge from a local defect. Plate 30, which was photographed in my laboratory by Dr Verma, is a highly complex pattern emerging from what must be a linear array of defects. This looks, on its tiny scale, like a mighty spiral nebula

PLATE 30 Complex of multiple spirals on a crystal of silicon carbide. The group originates from a region of a linear array of crystal defects

in the heavens. A closely related complex, also photographed by Dr Verma, is shown in Plate 31. Here the spiral groups emerge from an elliptical hole. In some crystals such a hole penetrates right through the crystal from side to side almost as if a fine hole had been drilled through. Magnification in Plates 30 and 31 is ×200.

Decoration

On certain crystals, not only those of silicon carbide, a strange effect has been observed which enhances the appearance of a spiral. It is called 'decoration'. Indeed it has been known for many years that if one breathes on a silicon carbide crystal the condensation of vapour seems preferentially to select the edges of any spiral present and makes it much more visible. On some occasions greasy impurity has a similar effect. For instance, leaving a crystal in contact with oily plasticine or gently smearing it with oil, or grease, can leave a fine-scale decoration on the ledges of a spiral however shallow it might be. The visibility of the shallow edge markedly improves, whilst at the same time there can often be seen an accumulation of very fine dots (impurity globules of a kind) on the very edge.

A fine example of this decoration is Plate 32 which shows at the same time a rare example of a beautifully formed 60° triangular spiral on silicon carbide (clearly closely related to the hexagonal symmetry). The fine dotted impurity decoration concentrating in both light and dark globules on the ledges (this is a matter of how the impurity is illuminated) is most noticeable. There are of course no fringes on this; it is just a direct microscope picture.

PLATE 31 (left) A complex of spirals on silicon carbide centred upon an elliptical hole around which growth has developed

PLATE 32 (below) Impurity decoration on the edges of an unusual triangular spiral on silicon carbide

PLATE 33 Dendritic leaf-like overgrowth on a flat face of a crystal of silicon carbide

Decoration on crystals is not only produced by imposed grease impurity. On occasions most remarkable decorative *overgrowths* appear firmly attached to crystals. We do not know what causes these in silicon carbide, but certainly we do find on this crystal some remarkable overgrowths. They are so unusual as to justify showing them. Plate 33 shows a most attractive dendritic leaf-like pattern and around these are small circular features rather reminiscent of snow flakes. On another crystal, Plate 34, these circular patterns turn up more vividly. It is easy to be convinced that these are oriented and even based on a hexagonal symmetry, as one would expect in the type of growth known as *epitaxial*, wherein one crystal grows on another, which imposes its symmetry on the former. There does also appear to be some sort of orientation in the decoration round the edges; it seems to run more or less perpendicular to the edge.

This concludes our review of crystal surfaces. The next section will describe applications of some technical interest, namely studies on the hardness of various materials. Technology is deeply concerned with wear, wear of machines, bearing surfaces, textiles and so on. Wear involves the rubbing of surfaces and thus surface microtopography is intimately associated with wear phenomena. When it is realised that wear and corrosion are perhaps the costliest of items in all industrial economy, it will be realised that wear and hardness study are of great economic importance.

PLATE 34 Overgrowth patterns on a hexagonal plate crystal of silicon carbide

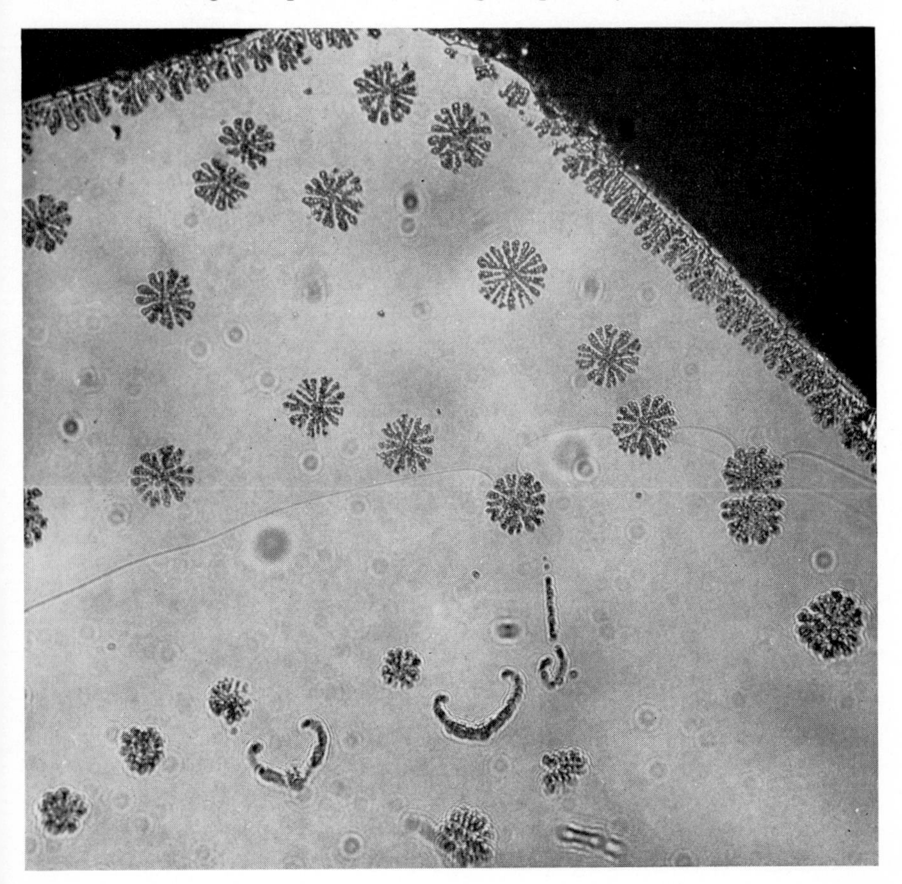

Chapter VIII

Hardness Testing

Measurement of hardness

Although hardness seems at first sight to have rather an obvious meaning, scientifically it is most difficult to measure, even to define. The technologist has been obliged to define a whole variety of different kinds of hardness. The earliest definition (that of Mohs) defines *scratch hardness*, a scale of 10 units being devised for minerals so that each material in the scale can scratch all those below it. Another hardness type was later recognised, i.e. resistance to wear or *abrasion hardness*. A third type, *indentation hardness*, then became popular as a measure of how far an indentor will penetrate into an object under a standard load. Technologically, this is now the most favoured kind of measure. Then again there is *rebound hardness*, i.e. how high will a hard ball rebound when dropped on the surface.

The most widely employed test for the engineer is the indentation method and yet even here there are several different hardness scales according to the *shape* of the indentor used. Generally, this shape is either a square-based pyramid (Vickers test) or an elongated pyramid (Knoop test), a shallow cone (Rockwell test) or a ball (Brinell test); other shapes have also been advocated. We restrict ourselves here to the widely used Vickers diamond pyramid test. A small square-based diamond pyramid is pressed, under a given load, for a standardised time, into the metal whose hardness is to be measured. If it is a hard metal the pyramid point penetrates only very slightly and thus makes a small square-shaped indentation. If the material is soft it goes in further and makes a bigger indentation. By measuring the area of the indent the load in kilograms per square mm which is being borne can be calculated and this is called the Vickers hardness. It is a valuable figure to the engineer.

Now it is clear that when the indentor goes in it must push some material away and thus around the indent the surface exhibits a certain amount of flow.

The indent need not be very large, maybe 1/5 mm across, so in any case it needs to be measured with a microscope. Yet small, or large, material must be pushed around somewhere by the indentor and it is of much interest to know where, and by how much. This is where interferometry comes into the story, for if we first polish a metal surface reasonably smooth, indent it and then view the surface interferometrically, we can anticipate being able to find out whether material has been pushed round by the indentor and if so where and by how much. The original smooth flat surface topography can be expected to be distorted.

Pyramid indentations

Plate 35 ($\times$ 100) is an interferogram of a Vickers indentation made on a steel plate using a load of 5 kg. We see a neat square hole and the fringes show that the metal has piled up around the hole. The fringes also reveal a neat point for they indicate that the indentor has not gone in quite truly to the perpendicular, since on two adjacent sides the pile-up is three fringes whereas on the other two it is four fringes. Note that surface disturbance is a minimum at the corners and a maximum opposite the centres of the sides of the square indent. Traditionally engineers have measured the *diagonals* of the indent and we see how wise they were.

On increasing the load by eight times to 40 kg, the indent and interference pattern of Plate 36 is secured. Pile-up has increased variously now to 11 and 13 fringes opposite the faces, but is still a good deal less (5 and 6 fringes) opposite the corners.

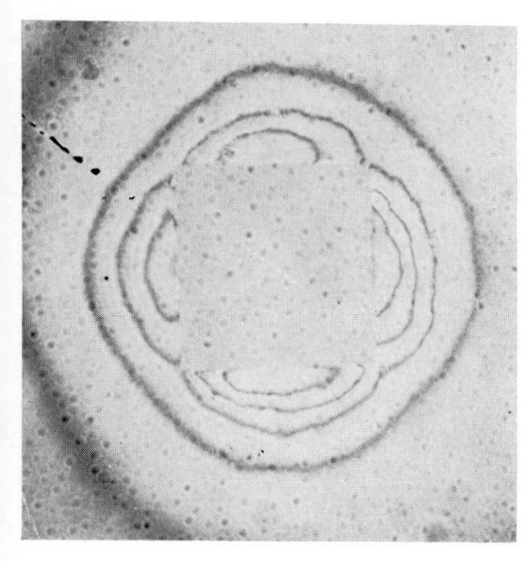

PLATE 35 Interferogram revealing the slight pile-up of metal surrounding the square-shaped indent made by a pyramid penetrating into steel

PLATE 36 Increased distortion over Plate 35 produced when a higher indenting load is used

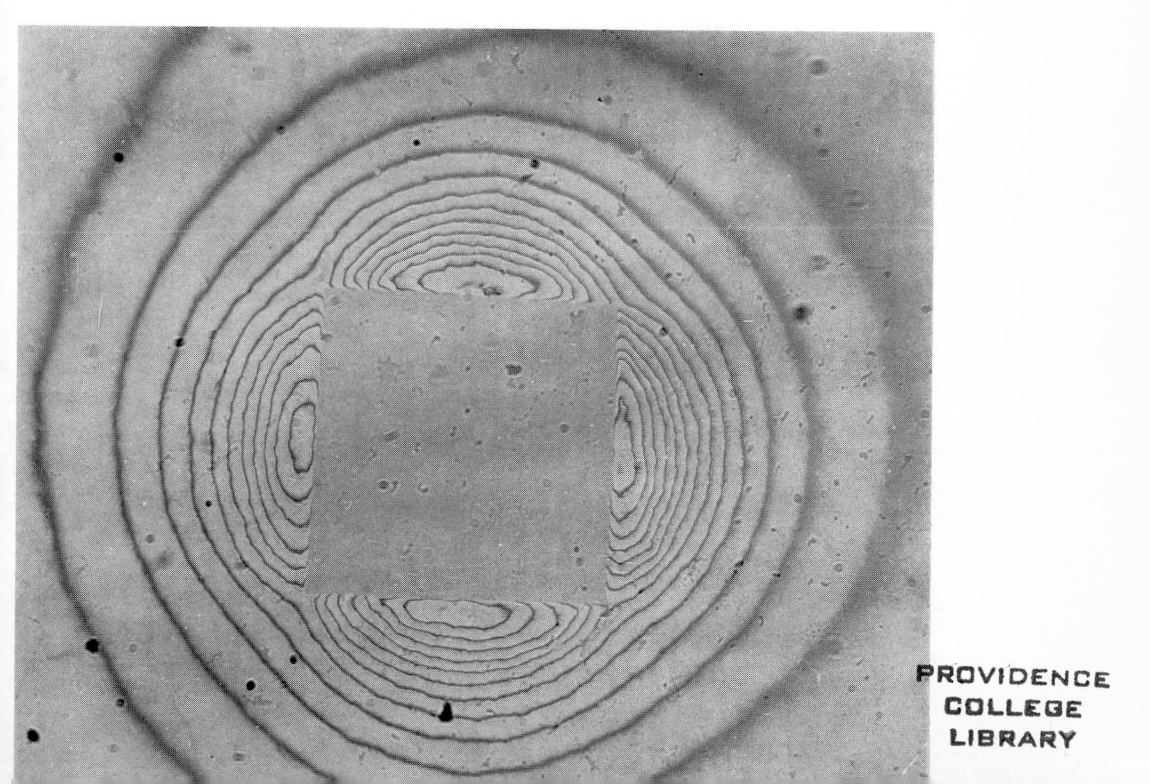

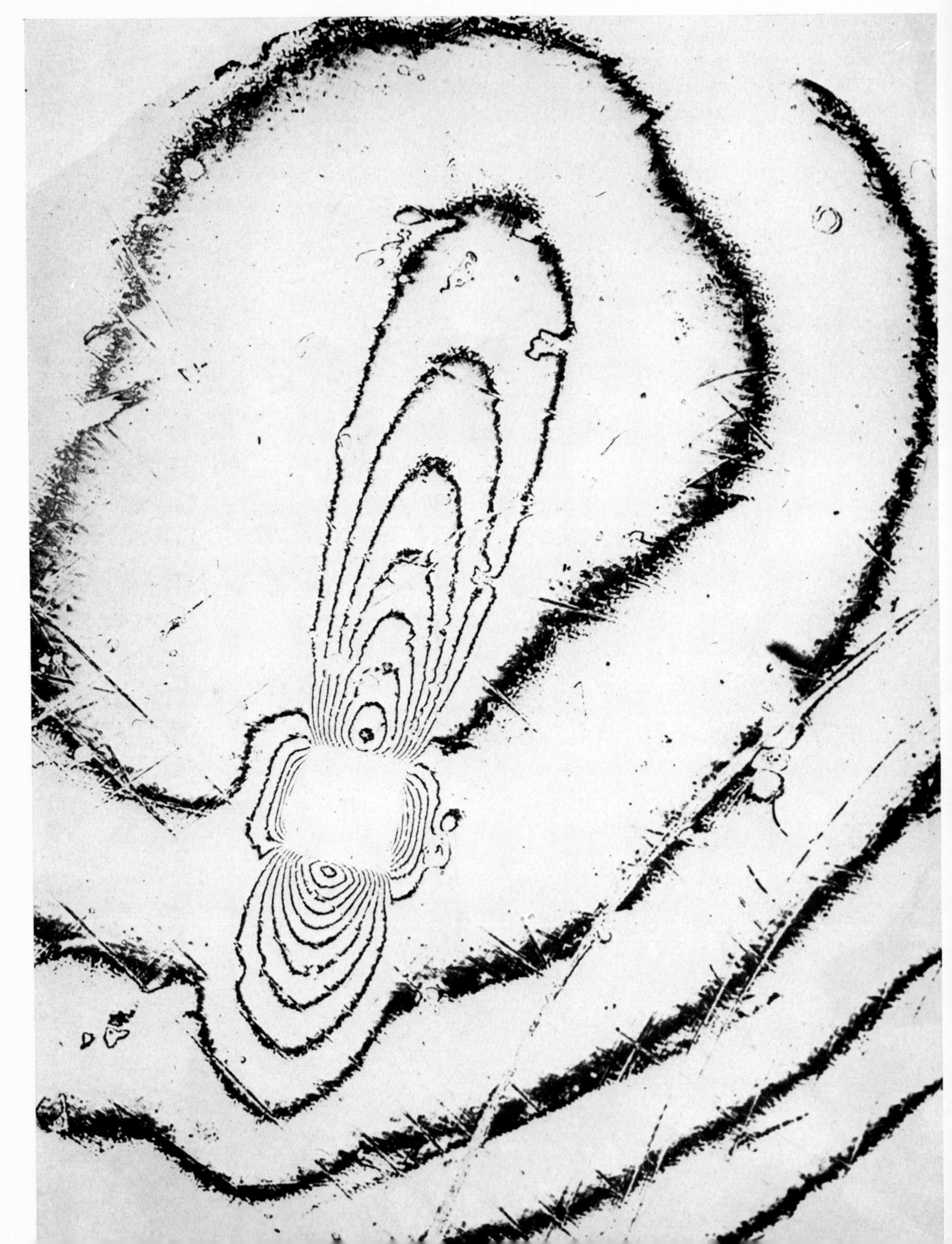

PLATE 37 Fringe pattern of distortion surrounding an indentation made into a single crystal of tin. Crystallographically directional hardness variations are shown

The pattern is practically symmetrical because, for an indent of this size, so small are the separate crystallites which form the structural units in steel that we can regard the material as being effectively uniform. Now how very different is the result obtained when we make an indent on a piece of metal which has been carefully grown to be in the form of a single crystal. Crystals we know to have directional properties and they differ a little in hardness in different directions. Yet, despite knowing this, the flow patterns we find for Vickers indents on single crystals of metal are most astonishingly unsymmetrical. Plate 37 is the interferogram given by a cast single crystal of tin. Its surface was not polished since polishing will influence surface hardness. The crystal was grown by casting against a piece of glass. A single crystal formed this way, when indented, gives a pattern strikingly notable in the directional asymmetry of the surface pile-up created by the pyramid indent.

Furthermore we have been able to demonstrate that the direction in which the large wings lie, is in fact a specific crystal direction (it is what the crystallographer calls the *c*-axis). A picture like this can offer guidance to an engineer when testing hardness of small crystalline units in alloys: it warns him that distorting effects produced by the indentor itself extend much further in crystallites than in uniform material and he must guard against this. This information was first revealed by this interferogram and by others like it obtained with other metals.

Directional hardness

It is particularly desirable to know the hardness of diamond and to try to measure how this is dependent on crystal direction. It is however extremely difficult to measure diamond hardness. The hardness differs in different directions on the crystal, and from the viewpoint of polishing and fabrication it is necessary to secure some measure at least of the relative hardness in the different directions. We have made such measurements by the following method. A small wheel has its edge shaped to a conically pointed rim which is then impregnated with diamond powder. The wheel is rotated at high speed and under a standard load and for a fixed number of turns it is applied to a diamond, in a selected direction. A small rut is ground into the diamond, maybe a millimetre or less in length. By examining such a rut with precision interferometry, despite its small size, one can see both its shape and depth and evaluate the volume of material ground away.

By repeating this in different directions, the different volumes abraded can be assessed and so an abrasion hardness can be estimated for the various directions selected.

Plate 38 ($\times 300$) (overleaf)shows an interferogram given by a small rut cut into a diamond. A computation can be made of the volume of material removed. When this was repeated in different directions a useful measure of differential hardness was obtained from the different cuts in the different directions. Diamond polishers have known for centuries about the different hardnesses in different directions, but only by such experimental tests can long-held impressions be put on a numerical basis.

PLATE 38 Fringe pattern of a rut one millimetre long ground on to a plane diamond surface

Chapter IX

Impact at High Speed

Collision with a drop

It has been known for some years that fast aircraft, missiles or rockets, when travelling at very high speeds through rainstorms, suffer erosion damage through impact with the raindrops. Measures of the gross amount of damage in heavy storms have been made. It became a matter of interest to attempt to assess the damage done by impact with a *single* drop. This we succeeded in doing with the

PLATE 39 Interference pattern of the depression made on a Perspex projectile striking a water droplet when it impacts at a speed of 750 miles an hour

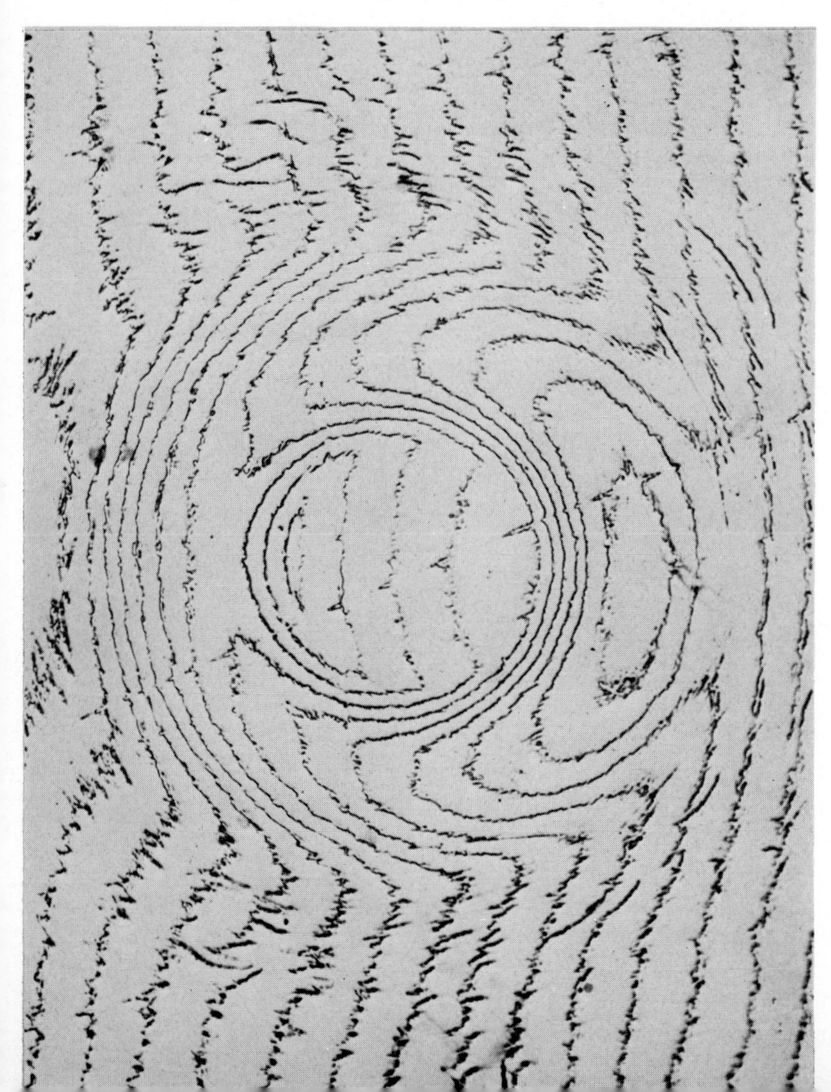

help of interferometry. The experiments were arranged in the following way. A drop of water of radius about 1 mm was suspended from a fine fibre. This was to be the single raindrop. A compressed air gun was designed which would fire at the drop small polished cylinders of, say, metal alloy, Perspex and so on, with speeds up to 1000 miles an hour. The projectile was thus to simulate a fast aircraft striking a single raindrop. Velocities were measured through the projectile intercepting two separated beams of light which illuminated photocells. The short time interval between occlusions of the light beams was determined electronically.

After striking the raindrop the projectile carried on and when spent was caught in a large box filled with cotton wool.

Some subsidiary tests showed that this process of collection did not produce any deleterious effects. The projectile was silvered on the impacted surface and then examined interferometrically. The mass of the projectile was sufficiently bigger than that of the droplet to eliminate any serious momentum transfer effects. Finally, the damage by impact was studied.

The effect of an impact at 750 miles an hour on Perspex striking a water drop is shown in Plate 39 ($\times$ 150). The water drop must have flattened somewhat since the pattern has the following meaning. Over a disc of diameter about 1 mm we see a moon-like crater, a raised lip, with a flat-top, cone-like hill in the centre of this. Surrounding this is a region of cracking and crazing. At higher velocity the damaged cracked region is much more extensive. Plate 40 (at $\times$ 60 only) shows the extensive crazing at 1000 miles per hour. Such studies enabled us to find a relation roughly between the volume of

PLATE 40 Fringe pattern showing severe crazing on Perspex when striking a water drop at 1000 miles an hour

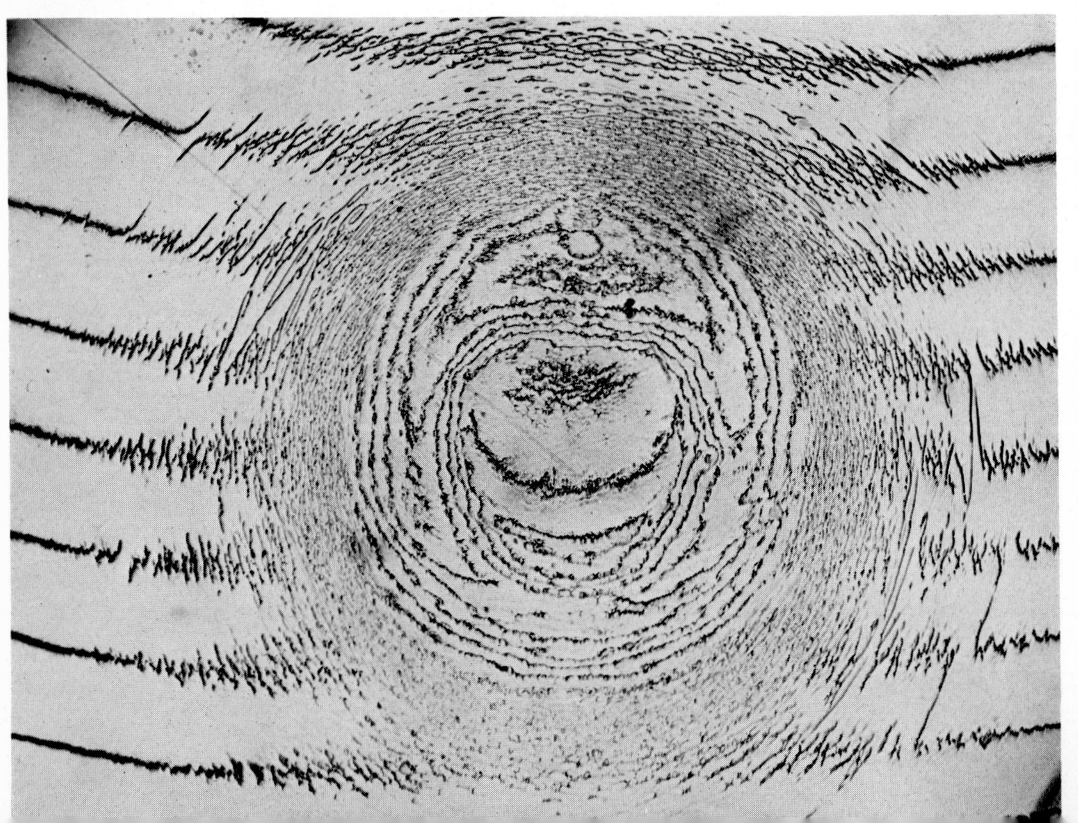

material eroded at one impact and the velocity V. It was found that there was an extremely high power-law, the damage being more or less proportional to the eighth power, i.e. to V^8.

This enormously high power-law explains why the amount of damage increases at a furious rate with speed, as is actually found in aircraft. For, if there be a certain amount of damage at 500 miles per hour, at 1000 miles it is 256 times as much, and at 2000 miles per hour it is actually over 65,000 times as great as it was at 500 miles an hour! It is not surprising that even *metal* fabric can and does suffer serious damage with prolonged raindrop collision at really high speeds. Fortunately at very high speeds aircraft and projectiles travel above rain cloud.

The impact on a hard aluminium alloy projectile (Duralumin) of a single water droplet, though less spectacular in its effect, is still notable. At about 700 miles per hour impact the hollow pit shown by Plate 41 ($\times 150$) appears. Here we see that the droplet, although having perhaps flattened, still behaves as a pseudo-solid sphere. It has not broken up and has produced a clean, spherical, cap-like indent, some ten fringes in depth and a little less than 1 mm across. The drop must have flattened a good deal, yet its penetrating power into the metal is surprisingly high.

PLATE 41 Interference pattern showing the circular pit on a Duralumin object striking a water drop at a speed of 750 miles an hour

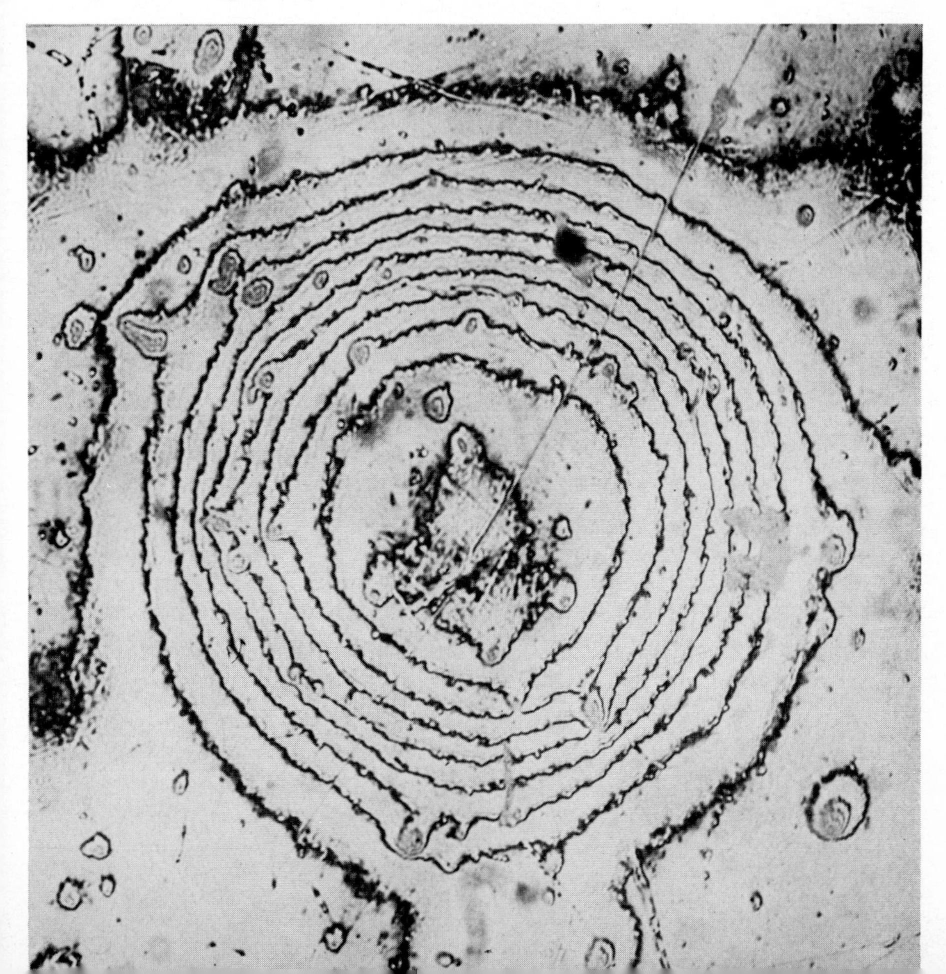

Chapter X

Some Technical Surfaces

Machined metals

It is of interest to find out just how well metals can be finished and polished by mechanical means. There exist 'superfine' turning methods which use diamond cutting tools and hones and it is considered that on hard steels a very superior finish can be obtained, especially on cylinders. When a theoretically perfect cylinder is examined with interference fringes, a series of straight line fringes can be obtained which run symmetrically along the cylinder and progressively get nearer because of the curvature. Deviations from perfection will readily be recognisable.

Plate 42 shows the fringe pattern given by a *superfine finished* steel rod. The finish is surprisingly good. Irregularities in general are only a small fraction of the fringe-order separation. A quite small

PLATE 42 Fringe pattern given by a steel cylinder which is commercially classed as of 'superfine' finish

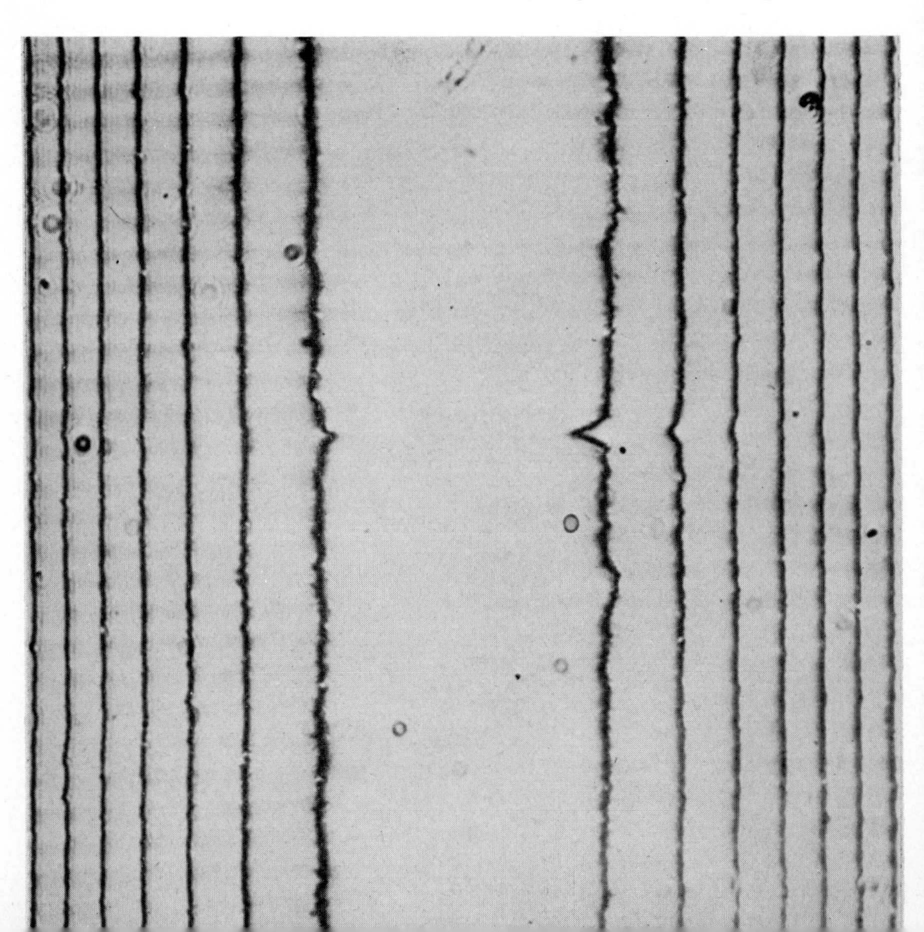

scratch is noticeable. Near the centre of the pattern the dispersion increases, so that the scratch is not really as deep as it looks at first sight.

Now, very good as this surface is, it is instructive to compare it with the fringe pattern given by a piece of ordinary, cheap, drawn, glass tubing. Since the tube used was of smaller diameter than the metal rod the fringe pattern is on a smaller scale; light-on-dark fringes can now be obtained in transmission since we are dealing with a transparent object. The superb smooth fringe pattern given by this simple untreated surface is shown in Plate 43. The fringes are quite straight and entirely free from any microstructure. Clearly the surface has an extraordinary degree of smoothness, which is due simply to the fire-polishing undergone in manufacture. It is because of fringe patterns like this that we can have complete confidence in the exploitation of fire-polished glass as a smooth reference surface for studying microtopographies.

Low reflectivity multiple-beam fringes

In many of the examples shown so far the reflectivities aimed at on the two matched surfaces have been as high as possible. A number of metals exist with a fairly high natural reflectivity of their own, in which case it is possible to obtain reasonably good fringe definition without the necessity of coating the object with silver. If the metal reflectivity is of the order of say 80 per cent, then good fringes are obtainable by matching against a surface of high reflectivity, such as say 90 per cent. We have in fact already illustrated fringes produced in this particular fashion from a natural tin crystal (unsilvered) which has an inherent reflectivity of perhaps 82 per cent. Plate 37, showing the indent on the

PLATE 43 The smooth structure-free fringe pattern shown by ordinary fire-drawn commercial glass tubing

tin crystal, was secured this way and it will be noted that the fringes, although adequately sharp for the purpose, are very much broader than, for instance, the extremely narrow high definition fringes of Plates 6 and 7.

When the object has a still lower natural reflectivity, say 70 per cent or so, it is still possible to get good definition by matching with a glass flat silvered with a more or less comparable reflectivity. There is true multiple-beam interference, but the lower reflectivities produce fringes which are a good deal broader than those obtained under very high reflectivity conditions. Nevertheless much of value is often secured. Plate 44 shows an interferogram obtained by matching an etched germanium crystal, which was unsilvered, against a suitably silvered flat. Germanium is a shiny metal and the reflectivity of the specimen was between 60 and 70 per cent. The fringe definition is quite good and the character and sizes of the various slightly elliptical etch hollows can be studied in detail from the interferogram.

There are rare occasions on which a metal surface might cause an embarrassing alloy if treated with silver and in such conditions, provided the natural reflectivity is adequate, quite good fringe definition can be secured.

PLATE 44 An etched germanium crystal face

Two-beam interference

All the pictures we have shown so far have been produced by *multiple-beam* interference methods. Excluding the relatively rare cases where the object to be studied has an inherent high reflectivity, the use of multiple beams implies that the object under study should be coated in vacuo with a layer of silver. Now clearly not every object can withstand the rigour of being kept in a vacuum. Innumerable biological objects for instance could not withstand a vacuum.

There is thus an extensive field of application of interferometry which uses *only two* interference beams instead of multiple beams. The fringes so formed, although broad and insensitive compared with multiple-beam fringes, are by no means to be despised, for they can give much valuable information. To show that this is so, we select here two examples where multiple-beam methods are *not* feasible. The first example concerns painted surfaces, the second shows unexpected structure on the surface of a tomato.

Painted surfaces

We shall illustrate the use of two-beam interferometry by looking at an ordinary domestic surface, a surface painted with ordinary glossy finish household paint. If such a finished surface be introduced into a vacuum to receive a silver deposit for multiple-beam purposes, it can be expected that the degassing from the paint film might produce changes in the shape and character of the surface coat. The object of this experiment was to find how the rate of drying of the paint affects the final finish. The finish was assessed interferometrically.

Two identical flat discs were painted at the same time. One was allowed to dry *slowly* in air, the other was dried *fairly quickly* with a current of mildly warmed air. Plate 45 ($\times$ 10) shows a surface after normal drying. The paint surface is surprisingly smooth and nearly flat. The fringes are of

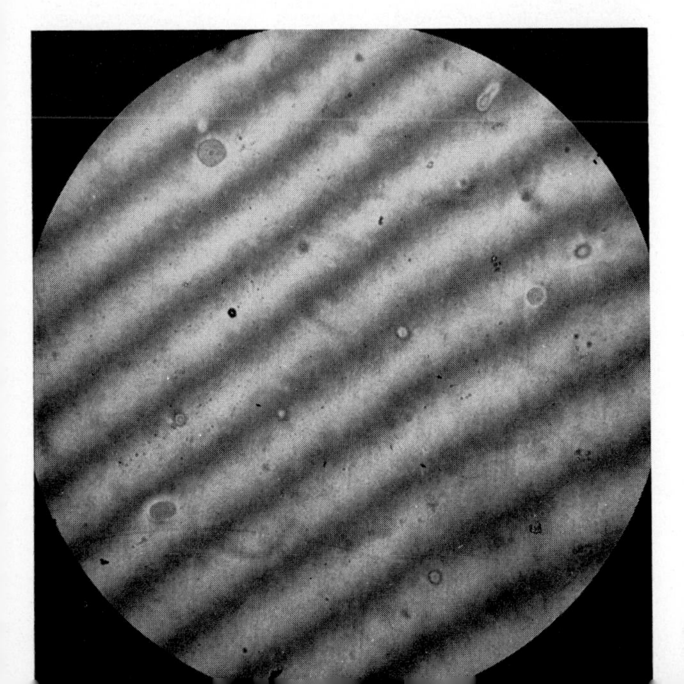

PLATE 45 Two beam fringes shown by a painted surface, the paint having dried slowly

course broad (two-beams) and cannot be expected to show up the microtopography which multiple beams can reveal, yet they do show a slight kink. On the whole the surface is good. Now, on the contrary, the effect of the quick warm air drying is very marked, as shown in Plate 46. It is apparent that a fierce wrinkling into hollows and ridges has taken place. It is clear that the relatively insensitive two-beam fringes are more than adequate for the purpose of showing the surface condition.

This wrinkling shows most clearly why it is that a rapidly dried painted surface has a slightly mottled appearance.

Surface of a tomato

Either a green or a ripe red tomato has a reasonably shiny surface, yet usually there exists a slight dappling or mottling, and at first impression such a surface does not look highly specular. Examination by interferometry of such a surface at $\times 500$, using a 4-mm microscope objective, gives an

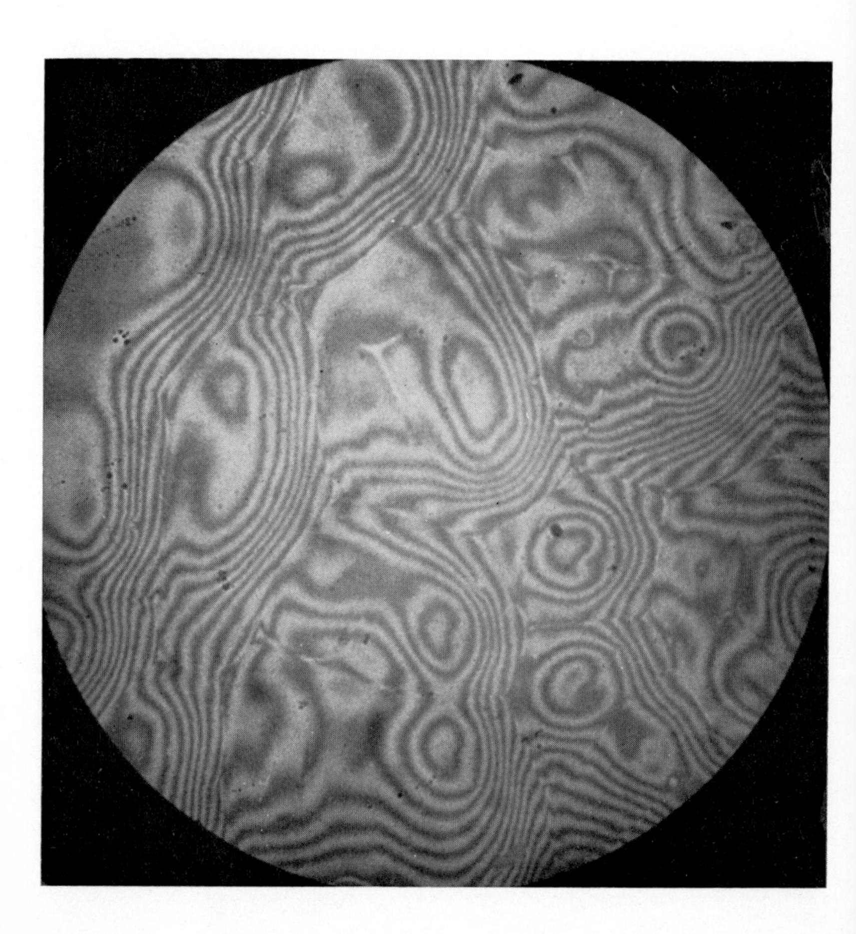

PLATE 46 The fringes from an identically painted surface to that on Plate 45, but here the paint has been rapidly dried with a current of warm air

astonishing result. It shows that the surface of a tomato is highly specular indeed; Plate 47 is a striking *two-beam interferogram* which is typical. The tomato has been placed upon a selected thin cover slip and interference fringes formed between the two surfaces in contact. In some tomatoes the surface consists of notable groups of hillocks, in others there are circular depressions. It is fairly easy to tell which is which by looking at the fringes formed by white light. With hillocks, coloured fringes appear at the hill tops, with valleys, they appear round the outer upper rims. Plate 48 shows a group of hillocks.

Some tomatoes start off showing hillocks but as they dry out the shrinkage converts these into valleys. We have only just developed this technique and are pursuing it vigorously to a variety of biological materials. Clearly two-beam interferometry can have very useful applications, even if the sensitivity is at best only a fiftieth of that of multiple-beam methods.

PLATE 47 (bottom left) Fringes given by the surface of a tomato

PLATE 48 (bottom right) A grouped set of hillocks shown by the surface of a tomato

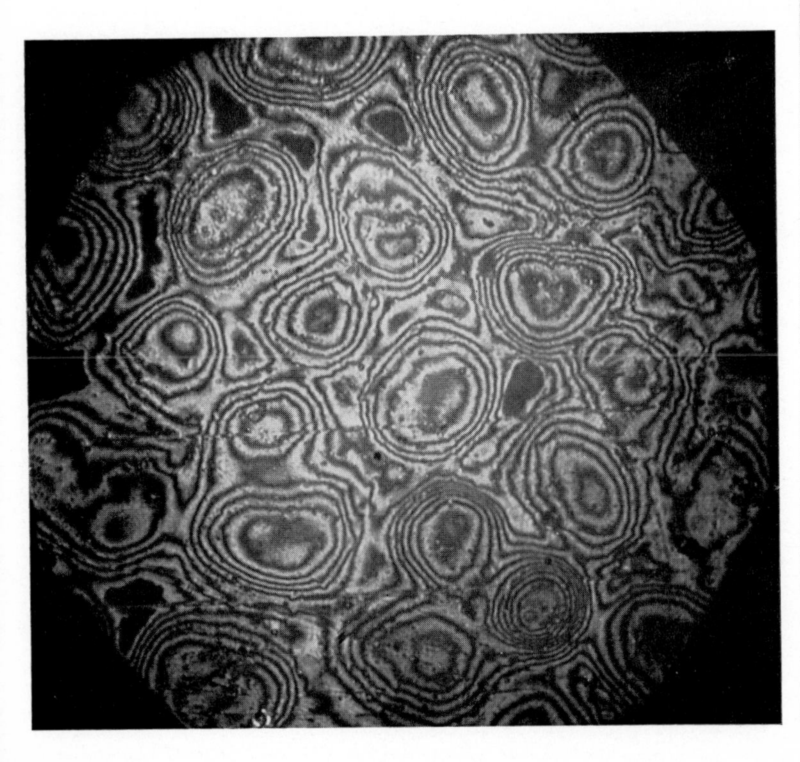

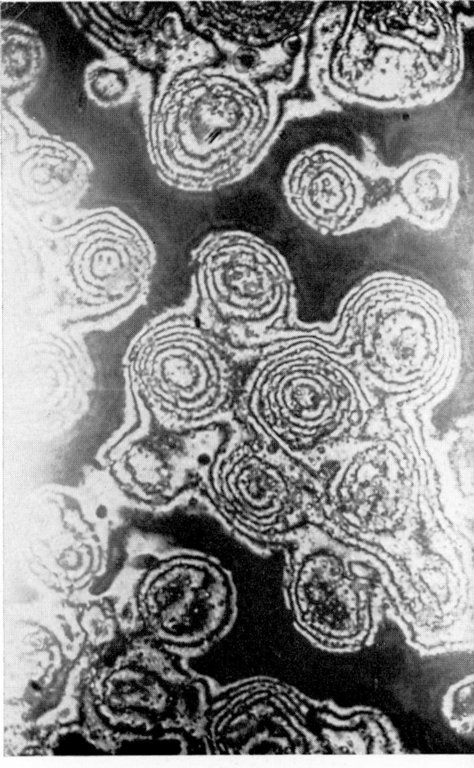

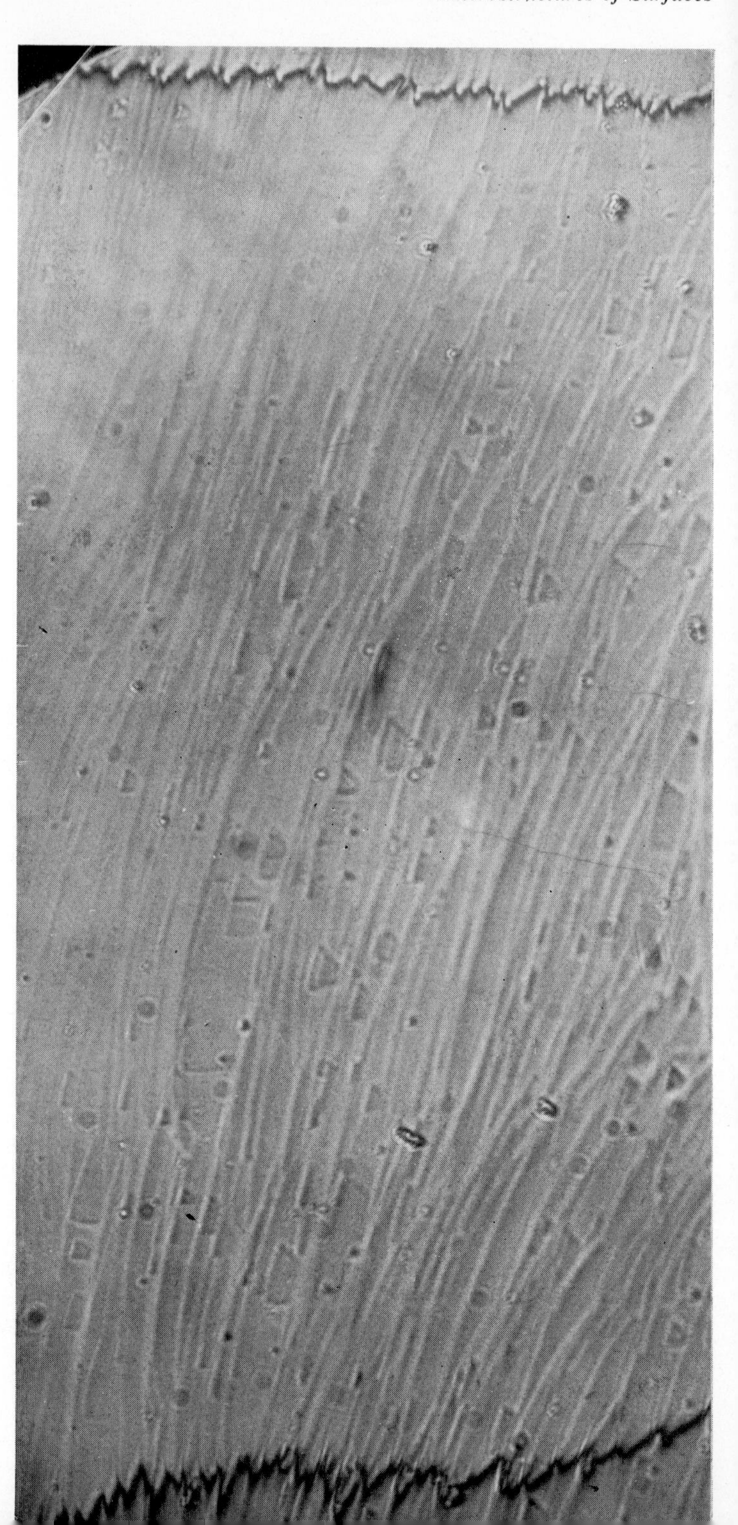

Conclusion

We have seen that interferometry, especially when refined by multiple-beam methods is a powerful weapon for unmasking the subtle microtopographies which exist both on natural surfaces and on those exposed to some treatment or other. The study of surface microtopography is beginning to be recognised by industry and technology as of fundamental significance. After all, one of the greatest costs of human economy is *wear*. Wear of engine parts, wear of shoes, wear of clothes, wear of tyres and so on. Since wear is always due to the rubbing of surfaces it is clearly of importance to find out what those rubbing surfaces really do look like. Engineering is only slowly beginning to appreciate what interferometry can offer here. Another field of great wastage to human economy is produced by *corrosion and rusting*. Great efforts and enormous expense go into protection of surfaces either by tinning, zincing, plating, painting or lacquering. But corrosion problems and failures in protection are again dependent on surface microtopography characteristics, so here too interferometry is beginning (but only too slowly!) to be used as an investigating tool.

Although this is a technique costing little other than skill and 'know-how', when used at its best the resolution and magnification (in one direction only, of course) equal, indeed perhaps exceed, the corresponding limits of the most sophisticated and costly of electron microscopes. The two systems are not in competition, but rather in certain instances can complement each other. Indeed their fields of application are different.

We shall conclude by showing one outstanding fringe picture in which dispersion is on an exceptionally large scale, yet a surprisingly sharp definition is marked. In Plate 49 we show two fringes at high magnification on the surface of a synthetic crystal of quartz. On the original print the fringes were 28 cm apart. The magnification thus has the formidable value of 1,000,000. True there is some empty magnification, yet the fringe definition remains surprisingly good. It can be seen precisely how the fringe zigzags correlate closely with surface markings. Also of note is that a half a millimetre displacement on these fringes, which is reasonably safely recognised, corresponds almost exactly to five angstrom units.

All the pictures shown in this text have been taken in the author's own laboratory, almost all of them by himself. Some were taken for him by students working under his direction. A number have appeared in the book *Surface Microtopography*, S. Tolansky (Longmans, 1960) where the reader can find full explicit details of the technique of multiple-beam interferometry, together with references to individual Ph.D. theses in which some of the pictures here illustrated have been discussed. Some silicon carbide pictures, were taken in my laboratory and appeared first in the Ph.D thesis of my former student A. R. Verma and subsequently in his book *Crystal Growth and Dislocations* (Butterworth, 1953).

Multiple-beam interference microscopes are now available commercially. Excellent instruments are produced by Varian in the U.S.A. and Multimi in Sweden.

PLATE 49 Very high dispersion, high magnification fringes on the surface of a crystal of synthetic quartz